ACHIEVE 100

Science

REVISION

Pauline Hannigan

The Publishers would like to thank the following for permission to reproduce copyright material.

Photo credits
Page 23 © photokitchen; page 24 © leksele

Acknowledgements

Rising Stars is grateful to the following schools who will be utilising Achieve to prepare their students for the National Tests: Chacewater Community Primary School, Cornwall; Coppice Primary School, Essex; Edgewood Primary School, Notts; Henwick Primary School, Eltham; Norwood Primary School, Southport; Sacred Heart Catholic Primary School, Manchester; Sunnyfields Primary School, Hendon; Tennyson Road Primary School, Luton.

ISBN: 978 1 78339 553 8

First published in 2015 by Rising Stars UK Ltd, part of Hodder Education, an Hachette UK Company

Carmelite House

50 Victoria Embankment

London EC4Y 0DZ

Reprinted 2015

www.risingstars-uk.com

Author: Pauline Hannigan

Series Editor: Ed Walsh

Educational Consultant: Shan Oswald

Accessibility Reviewer: Vivien Kilburn

Publishers: Kate Jamieson and Gillian Lindsey

Project Manager: Debbie Allen

Editorial: Jo Murray, Lynette Woodward, John Durkin, Fiona Leonard

Cover design: Burville-Riley Partnership

Illustrations by John Storey, Pen and Ink Book Company Ltd

Text design and typeset by the Pen and Ink Book Company Ltd

Printed by Craft Print Pte Limited, Singapore

A catalogue record for this title is available from the British Library.

Contents

Welcome to Achieve Key Stage 2 Science Revision Book 100

In this book you will find lots of practice and information to help you be successful in the Key Stage 2 Science sampling tests.

About the Key Stage 2 Science National Sampling Tests

Not all schools sit the Science sampling tests; a selection is chosen to use them. The tests will take place in the summer term in Year 6. They will be done in your school and will be marked by examiners – not by your teacher.

The tests are divided into three papers:

Paper b: Biology – 25 minutes (22 marks)
Paper c: Chemistry – 25 minutes (22 marks)
Paper p: Physics – 25 minutes (22 marks)

- In each test, there will be a mixture of question types, including multiple-choice, labelling diagrams, short responses such as one or two words, or longer responses where you need to describe an experiment or explain your answer.
- The number of marks will vary depending on how difficult the question is.
- Between 25 and 35 per cent of the questions will test your ability to 'work scientifically'. This means using your scientific understanding to plan or analyse investigations. The questions might be based around a picture or a description of an investigation and its results. You might have to suggest how something could be investigated. See the opposite page for more on working scientifically.

Test techniques

Before the tests

- Try to revise little and often, rather than in long sessions.
- Choose a time of day when you are not tired or hungry.
- Choose somewhere quiet so you can focus.
- Revise with a friend. You can encourage and learn from each other.
- Read the 'Top tips' throughout this book to remind you of important points in answering test questions.
- Make sure that you know what the words in the glossary mean.

During the tests

- READ THE QUESTION AND READ IT AGAIN.
- If you find a question difficult to answer, move on; you can always come back to it later.
- Always answer a multiple-choice question. If you really can't work out the answer, have a guess.
- Check to see how many marks a question is worth. Have you written enough to 'earn' those marks in your answer?

- Read the question again after you have answered it. Make sure you have given the correct number of answers within a question, e.g. 'Tick **two** boxes'.
- If you have any time left at the end, go back to the questions you have missed. If you really do not know the answers, make guesses.

Where to get help:

- Pages 8–29 practise Biology.
- Pages 30–45 practise Chemistry.
- Pages 46–75 practise Physics.
- Pages 77–81 provide the answers to the 'Let's practise' and 'Try this' questions.

Working scientifically

Working scientifically (WS) is an important skill; it requires you to use your scientific understanding to plan, carry out and analyse investigations.
The table below shows the different aspects of working scientifically and identifies the useful skills for each one. The strands marked with an asterisk (*) cannot be assessed or can only be partially assessed in the Key Stage tests because they rely on practical skills or experiences linked to school location.

Planning	Asking relevant questions and using different types of scientific enquiries to answer them * Planning different types of scientific enquiries to answer questions, including recognising and controlling variables where necessary
Carrying out	Setting up practical enquiries, and comparative and fair tests *
Measuring	Making systematic and careful observations, taking accurate measurements, using a range of equipment with increasing accuracy and precision, taking repeat readings when appropriate *
Recording	Gathering, recording, classifying and presenting data in a variety of ways to help in answering questions * Recording findings using scientific language, drawings, labelled diagrams, classification keys, scatter graphs, bar and line charts and tables
Concluding	Using results to draw simple conclusions, making predictions for new values, suggesting improvements and raising further questions Identifying differences, similarities or changes related to simple scientific ideas and processes Using straightforward scientific evidence to answer questions, to support findings or to refute ideas or arguments
Reporting	Reporting and presenting findings from enquiries, including oral and written explanations, displays or presentations of results and conclusions, causal relationships and explanations of and degree of trust in results, in oral and written forms such as displays and other presentations *
Further work	Using test results to make predictions to set up further comparative and fair tests

In this book, you will find lots of opportunities to practise your WS skills in different ways – look out for the icons above.

How to use this book

1 ***Introduction*** – tells you what you need to be able to do for each topic.

2 ***What you need to know*** – summarises the key information for the topic with the correct terminology. Words in bold are key words and those in lilac are also defined in the glossary at the back of the book.

3 ***Let's practise*** – a practice question is broken down in a step-by-step way to help you to understand how to approach answering a question and get the best marks that you can.

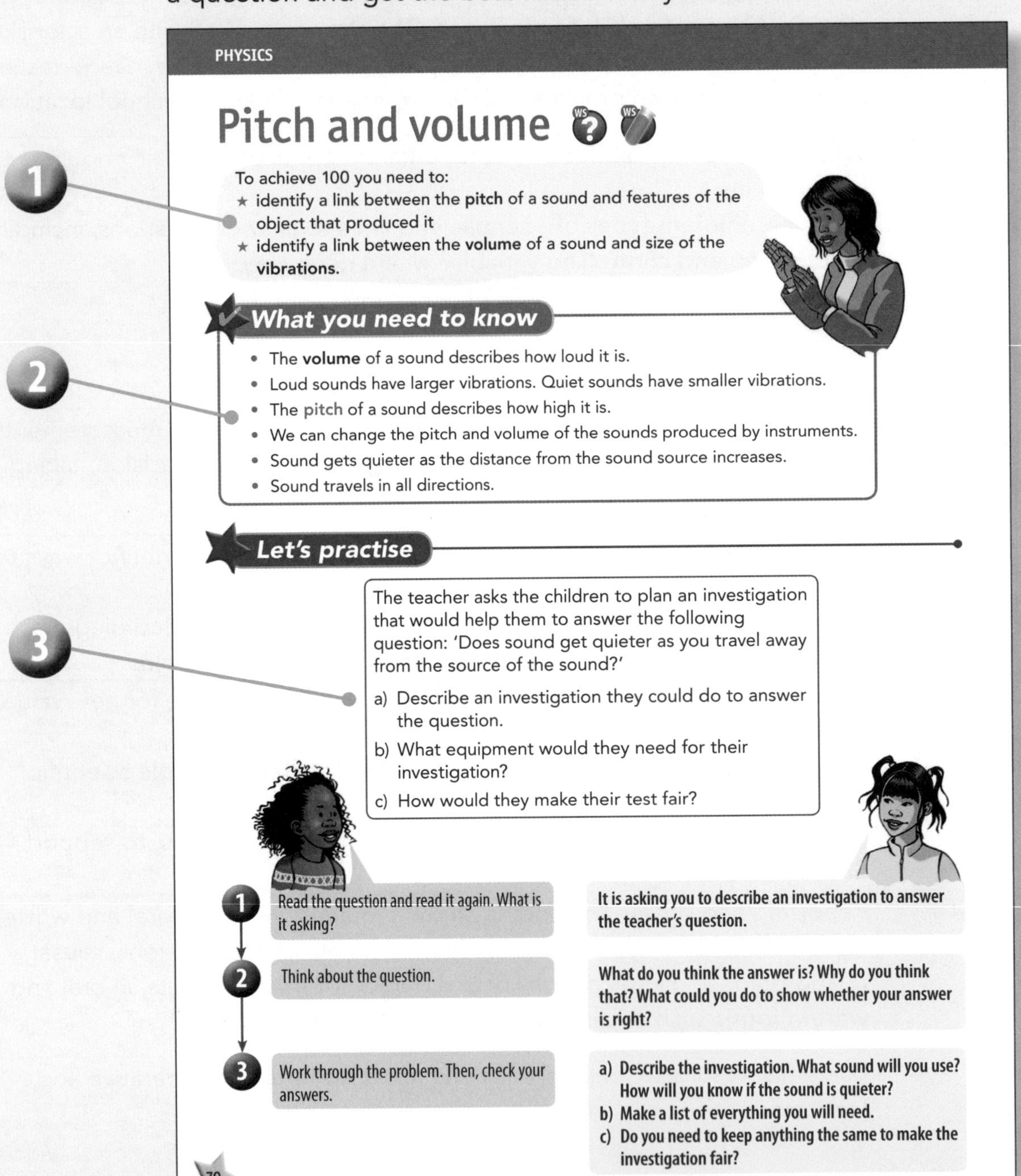

PHYSICS

Pitch and volume

To achieve 100 you need to:

- identify a link between the **pitch** of a sound and features of the object that produced it
- identify a link between the **volume** of a sound and size of the **vibrations.**

What you need to know

- The **volume** of a sound describes how loud it is.
- Loud sounds have larger vibrations. Quiet sounds have smaller vibrations.
- The pitch of a sound describes how high it is.
- We can change the pitch and volume of the sounds produced by instruments.
- Sound gets quieter as the distance from the sound source increases.
- Sound travels in all directions.

Let's practise

The teacher asks the children to plan an investigation that would help them to answer the following question: 'Does sound get quieter as you travel away from the source of the sound?'

a) Describe an investigation they could do to answer the question.

b) What equipment would they need for their investigation?

c) How would they make their test fair?

1. Read the question and read it again. What is it asking?
   It is asking you to describe an investigation to answer the teacher's question.
2. Think about the question.
   What do you think the answer is? Why do you think that? What could you do to show whether your answer is right?
3. Work through the problem. Then, check your answers.
   a) Describe the investigation. What sound will you use? How will you know if the sound is quieter?
   b) Make a list of everything you will need.
   c) Do you need to keep anything the same to make the investigation fair?

70

4 ***Try this*** – this is where you get the chance to answer questions for yourself by applying your scientific knowledge. There are a different number of questions for each topic.

5 ***Top tips*** – these give you further reminders about answering test questions or to help you understand a tricky topic.

4

PHYSICS

Try this

1 Describe **two** ways that you could make a higher note on a violin.

2 a) What do you think vibrates when a recorder is blown?

b) How does covering the holes on a recorder change the pitch of the note?

3 Describe how you could make the sound of a drum louder.

4 How could you show that the vibrations are greater when the sound is louder?

5

Top tip

- The more there is to vibrate, the lower the note.

71

Health and digestion

To achieve 100 you need to:

★ explain how humans and other animals need the right types of **nutrients** from their food
★ describe the function of parts of the **digestive system**
★ describe how diet, exercise, drugs and lifestyle can affect our bodies.

What happens to your food?

What you need to know

- Plants make the **nutrients** they need to live. Animals cannot do this so they have to eat food.
- Foods can be split into different food groups.
- A balanced diet is when you eat some food from all of the groups.
- Your diet can affect your health.

1 Let's practise

Choose the plate that has the most balanced meal. Give **two** reasons for your choice.

1. Read the question and read it again. What is it asking?
   The question is asking you to compare the two plates of food and explain your choice.
2. Look at the pictures.
   What food groups are shown on each plate?
3. Remember the key facts.
   Each food group has a different job to do. Decide how many groups are shown on each plate, and which one is more balanced.
4. Explain your choice.
   Give two reasons comparing the plates of food.
5. Check your answer.

Where does your food go?

What you need to know

- Food changes as it travels through the **digestive system**.
- The useful parts of food are moved round the body in the blood.
- The useless parts of food are removed from our bodies when we go to the toilet.

2 Let's practise

a) **Draw a line** from each word to the correct part of the digestive system.

b) Jack thinks digestion starts in the stomach. Abi thinks digestion ends in the stomach.

Do you agree or disagree? Explain why.

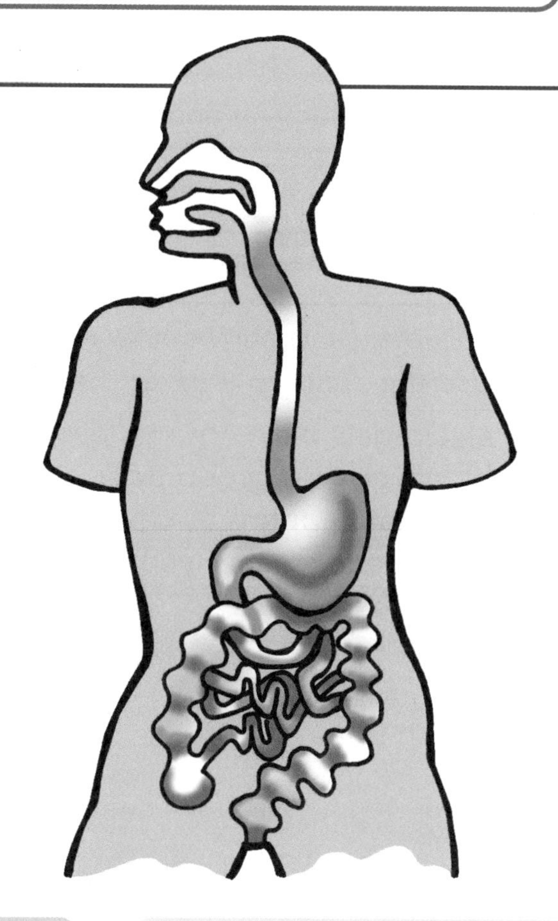

stomach

mouth

anus

oesophagus

small intestine

large intestine

1. Read the question and read it again. What is it asking?

   Part a) is asking you to match the names of the digestive system with the body parts in the diagram. Part b) is asking you to show your understanding of the digestive system.

2. Look at the diagram and where the parts are in your body.

   Think about the journey your food makes after you put it in your mouth.

3. Work through the problem.

   For part a), start with the names you know for certain. Think about what happens at each stage.

4. Use what you know to decide if the children are right or wrong.

   Where does digestion start? Where does it end? Use this information to answer part b).

5. Check your answer.

Looking after yourself 

What you need to know

- Exercise makes your **heart**, **lungs**, **muscles** and **bones** stronger.
- Eating a balanced diet keeps your body at a healthy weight.
- **Drugs** can affect the way you think and behave.
- **Smoking** and **alcohol** are addictive and have short-term and long-term health effects:
  - Smoking can damage your heart and lungs and may cause cancer.
  - Alcohol can damage your **liver** and **brain**.

3 Let's practise

This bar chart shows how much sport children play. Use the chart to answer these questions.

a) Do girls play more or less sport than boys?
b) Which year group does most sport?
c) What happens as children get older? What evidence makes you think that?

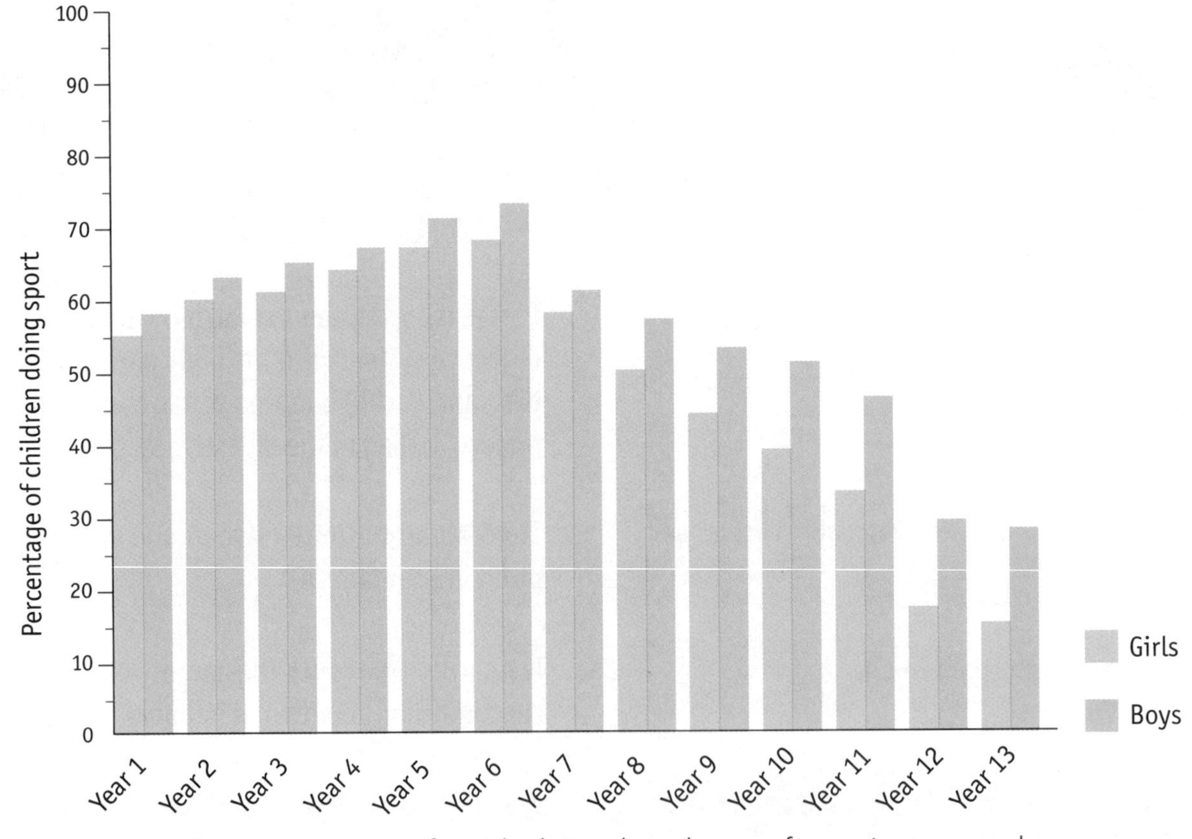

Bar chart to show percentage of pupils doing three hours of exercise per week

Read the question and read it again. What is it asking?

You are being asked to use the information in the bar chart.

Look for patterns in the way the bars go up or down.

The bars compare girls and boys in each year group. The chart also shows how the amount of sport changes through different year groups.

Work through the problem.

Check your answer.

a) Compare the two bars in each year group.
b) In which year group are the bars highest?
c) What happens to the bars as the children get older?

Try this

1 Why do you think young people do less sport as they get older?

____

2 Write **one** reason why it is important to exercise.

____

3 Why is it important to eat a balanced diet?

____

4 Write **two** ways you can keep your heart healthy.

____

____

Top tips

- In the test, you may be asked to explain what a bar chart shows.
- On a bar chart, the *y* (side) axis shows what is being measured. This axis should be labelled with the units.
- Always read the title of the bar chart carefully so you know exactly what it is about.
- Make sure you understand what the numbers on the *y* (side) axis are showing you. Make sure you know what the different groups are on the *x* (bottom) axis.
- Read the key if there is one.

Skeletons

To achieve 100 you need to:
★ describe the functions of the **skeleton** and **muscles**
★ identify some animals (**invertebrates**) that don't have skeletons.

What you need to know

- The skeleton supports the body, and protects the **organs**.
- Not all animals have a skeleton.
- The **joints** between some of your bones allow movement.
- Muscles are attached to your bones and allow you to move. They are made of an elastic material.

Let's practise

a) **Draw a line** from each word to the correct part of the skeleton.

b) Which **three** of these bones support the body?

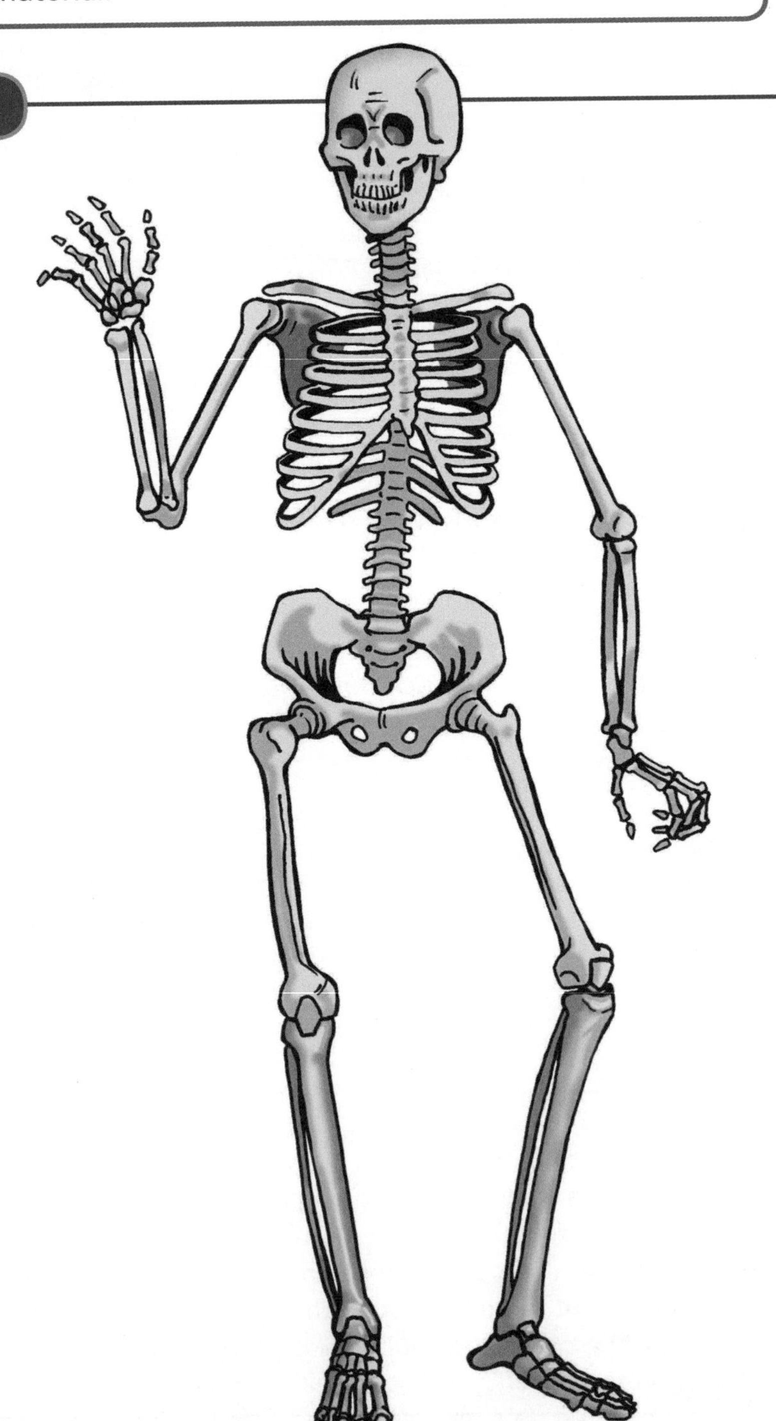

breastbone

thigh bone

pelvis

shoulder blade

spinal column

skull

ribs

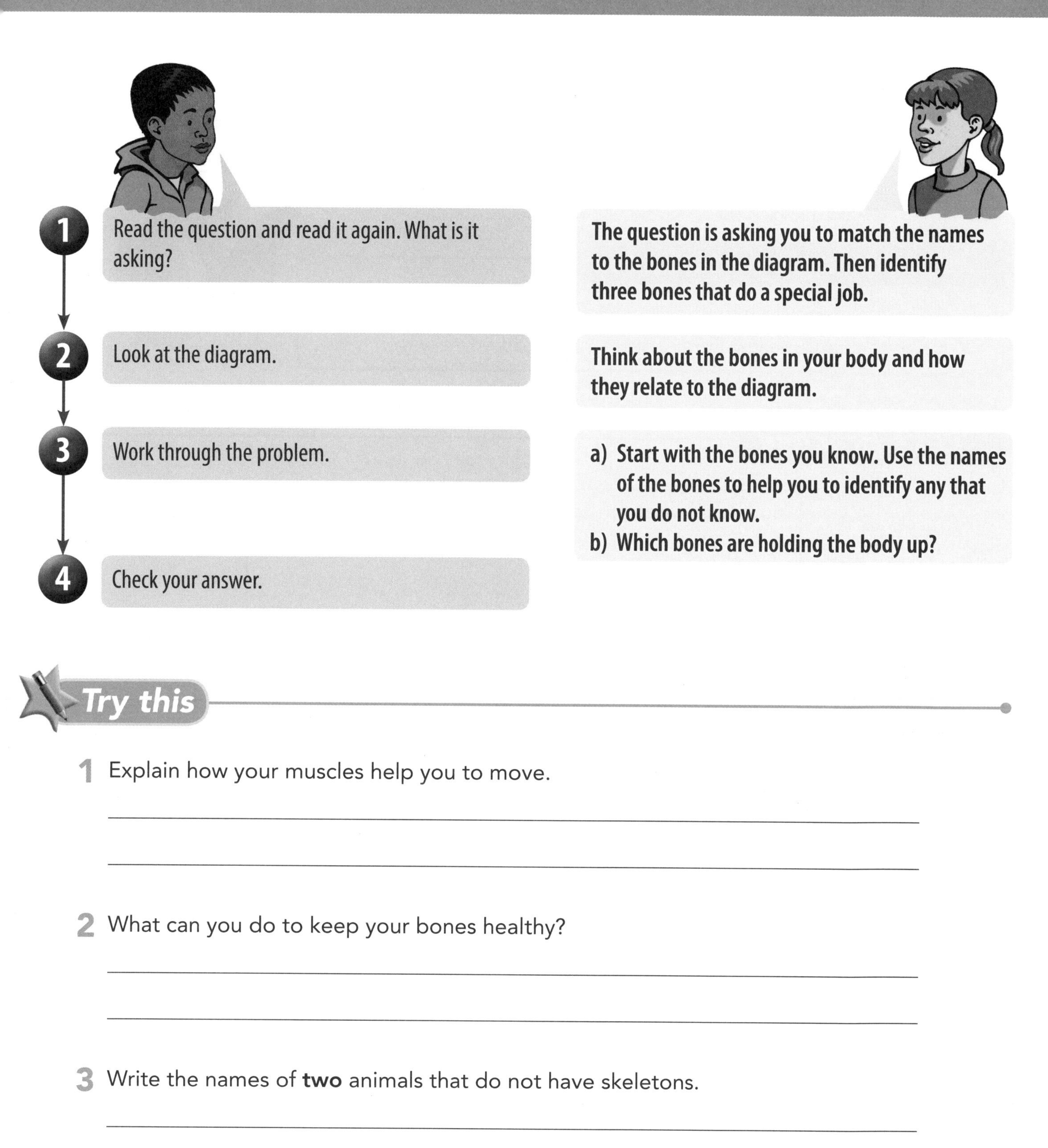

1. Read the question and read it again. What is it asking?

   The question is asking you to match the names to the bones in the diagram. Then identify three bones that do a special job.

2. Look at the diagram.

   Think about the bones in your body and how they relate to the diagram.

3. Work through the problem.

   a) Start with the bones you know. Use the names of the bones to help you to identify any that you do not know.
   b) Which bones are holding the body up?

4. Check your answer.

Try this

1 Explain how your muscles help you to move.

____

____

2 What can you do to keep your bones healthy?

____

____

3 Write the names of **two** animals that do not have skeletons.

____

____

Human development

To achieve 100 you need to:

★ describe the changes that occur as humans develop to old age.

What you need to know

- People change both physically and mentally as they get older.
- Changes happen at different times and in different ways.

Let's practise

Human life can be split into the following stages:

a) Infancy
b) Childhood
c) Puberty
d) Adulthood
e) Old age

Name a change that happens to a person at each stage.

1. Read the question and read it again. What is it asking?
   It is asking you to write about the changes that happen through life.
2. Remember the key facts.
   What are the approximate age ranges of each of the stages? What changes happen at each of these different stages?
3. Work through the problem.
   Think of yourself, your family or other people you know to help you to answer the question.
4. Check your answer.

Try this

As soon as you are born, you start to grow and you keep on growing.

1 Explain whether you agree with Amy.

____

____

Teeth

To achieve 100 you need to:

★ name the different types of teeth in humans and say what they do.

What you need to know

- Teeth vary in different animals according to what they eat.
- Different types of teeth are: **incisors**, **canines**, **premolars** and **molars**.
- Teeth, like bones, are alive.
- Hard **enamel** protects the tooth.
- Bacteria and sugar make **plaque**, which attacks the enamel.
- It is important to look after your teeth.

Let's practise

a) Label the four main types of teeth.

b) Join each tooth to its correct job (function).

For grinding and crushing food	For holding and tearing food	For grinding and crushing food	For cutting food

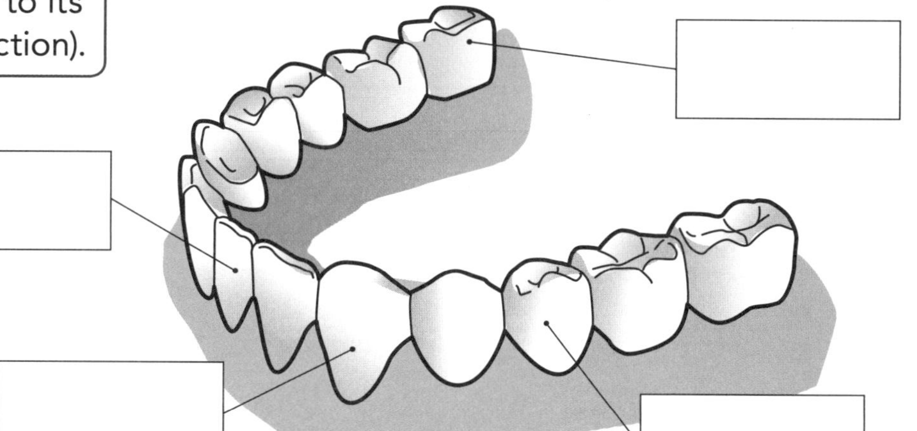

1 Read the question and read it again. What is it asking?

Part a) is asking you to label the diagram. In part b) you need to choose the correct use for each tooth.

2 Think about each part of the question.

Start with the names and functions you know.

3 Work through the problem. Then, check your answer.

Think about biting into an apple. Which teeth do what? Would it be the same if you were eating a chicken drumstick? Two types of teeth have the same job.

Try this

1 Write **two** ways of looking after your teeth.

____

____

Food chains

To achieve 100 you need to:

★ draw and interpret a **food chain**

★ name the **consumers, producers, predators** and **prey** in a food chain.

What you need to know

- Plants make the nutrients they need to survive from water, carbon dioxide and sunlight. Animals cannot do this and need to eat plants or other animals to survive.
- Plants are known as **producers** because they produce food for themselves. There is usually a green plant at the beginning of every **food chain**.
- Animals are known as **consumers**. They can be **predators** or **prey** (or both) depending on where they are in the food chain.
- Plants and animals are often part of more than one food chain.

Let's practise

Look at the food chain below. Write the word *producer, predator* or *prey* in each box. You can use more than one word in a box.

1 Read the question and read it again. What is it asking?

The question is asking you to label each part of the food chain. You might need to use more than one word for some animals.

Step		Hint
2	Remember the key facts.	**What does each word mean? How do they relate to each other?**
3	Work through the problem.	**What do most food chains start with? Think about what each animal eats AND what eats each animal.**
4	Check your answers.	**Start at the end of the food chain and work backwards. Think about what is on both sides of each animal.**

Try this

1 If only a few caterpillars hatch one year, how might this affect the food chain that the caterpillars are in?

____

____

2 Use the words below to make a food chain.

owl grasshopper rat grass

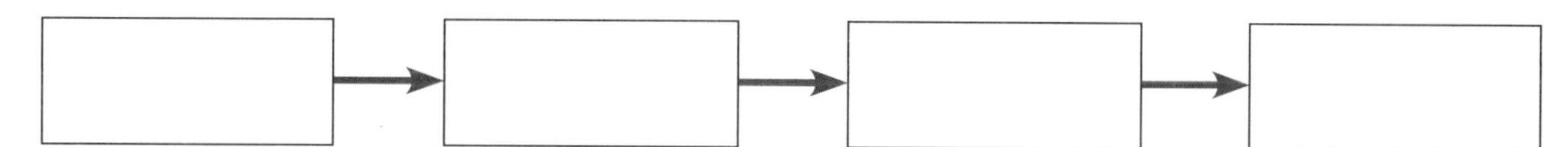

3 Draw a different food chain of your own.

4 Why is the producer the most important part of a food chain?

____

____

Top tip

- In a food chain, the arrows always go from left to right. Each arrow means *is eaten by*.

Heart and circulation

To achieve 100 you need to:

★ name the main parts of the human **circulatory system**
★ describe the **function** of the **heart**, **blood vessels** and **blood**
★ describe how **nutrients** and water are carried around the body.

Where does your blood go?

What you need to know

- The **circulatory system** is made up of the heart, the lungs, the **blood** and the **blood vessels**.
- Your body needs a constant supply of **oxygen** to keep it working.
- Your lungs take oxygen from the air and pass it to your blood, which carries it around your body.
- Carbon dioxide is carried in the blood to the lungs, then it leaves the lungs when you breathe out.
- Blood also carries the nutrients from food and water.

1 Let's practise

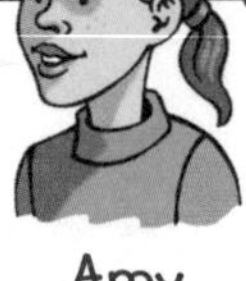

The heart pumps blood to the lungs to collect oxygen on the way to the rest of the body.

Amy

The heart pumps blood to the lungs to collect oxygen. The blood then comes back to the heart and is then pumped around the rest of the body.

Ben

The heart pumps blood to the body and on to the lungs to collect more oxygen. The blood then comes back to the lungs.

Sanjay

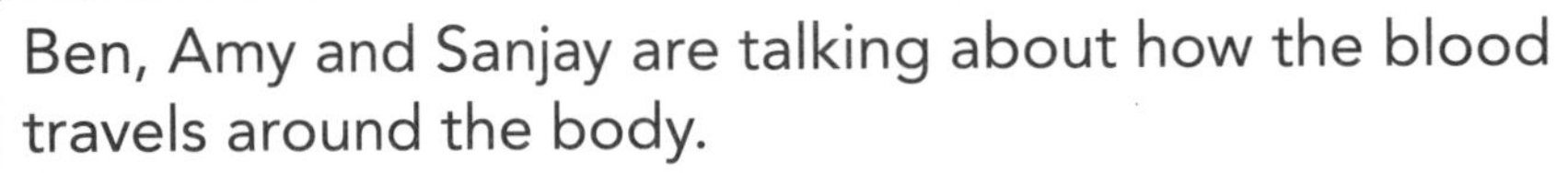

Ben, Amy and Sanjay are talking about how the blood travels around the body.

a) Who gives the best answer?

b) What mistakes are the other two children making?

1	Read the question and read it again. What is it asking?	**It is asking how the heart pumps blood around the body.**
2	Think about what each child is saying.	**Draw a diagram or sequence of steps for each child if you need to.**
3	Remember the key facts.	**How many times does the blood go through the heart in each full circuit? Who is correct? Which parts of the other statements are wrong?**

A beating heart 

What you need to know

- The heart beats faster when we exercise.

2 Let's practise

This line graph shows how a girl's heart rate changes from when she is resting, doing a warm up, exercising hard and then resting again.

a) For how long does the girl warm up?

b) What is the girl's heart rate after she has exercised?

c) Explain why her heart rate changes.

d) What evidence is there in the graph to show that people have a normal resting heart rate?

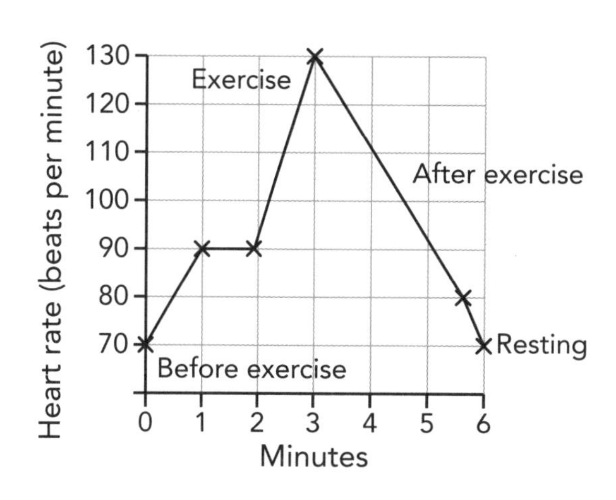

1 Read the question and read it again. What is it asking?

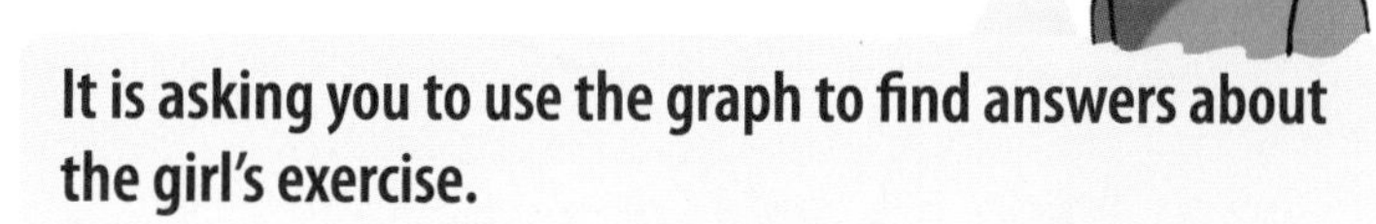

It is asking you to use the graph to find answers about the girl's exercise.

2 Look carefully at the graph.

Is the line going up? Down? Or staying the same?

3 Work through the questions. Then, check your answers.

a) Find the number of minutes on the *x* axis.

b) Find the heart rate in beats per minute on the *y* axis.

c) What happens when you exercise? What does your body need more of?

d) Compare the heart rate before and after exercise.

Try this

1 How does exercise help to keep your heart healthy?

____

2 Write **one** other way you can help to keep your heart healthy.

____

3 How do you measure your heart rate?

____

Classification and keys

To achieve 100 you need to:
- ★ describe the main groups in the **classification system** and ways that living things are grouped
- ★ use and make **keys** to help group, identify and name some living things
- ★ explain why scientists need to group plants and animals.

The key to identification

What you need to know

- **Keys** use questions to divide living things into groups.
- **Microorganisms** are living things. Some, like yeast, can be helpful. Some can be harmful and cause disease.

1 Let's practise

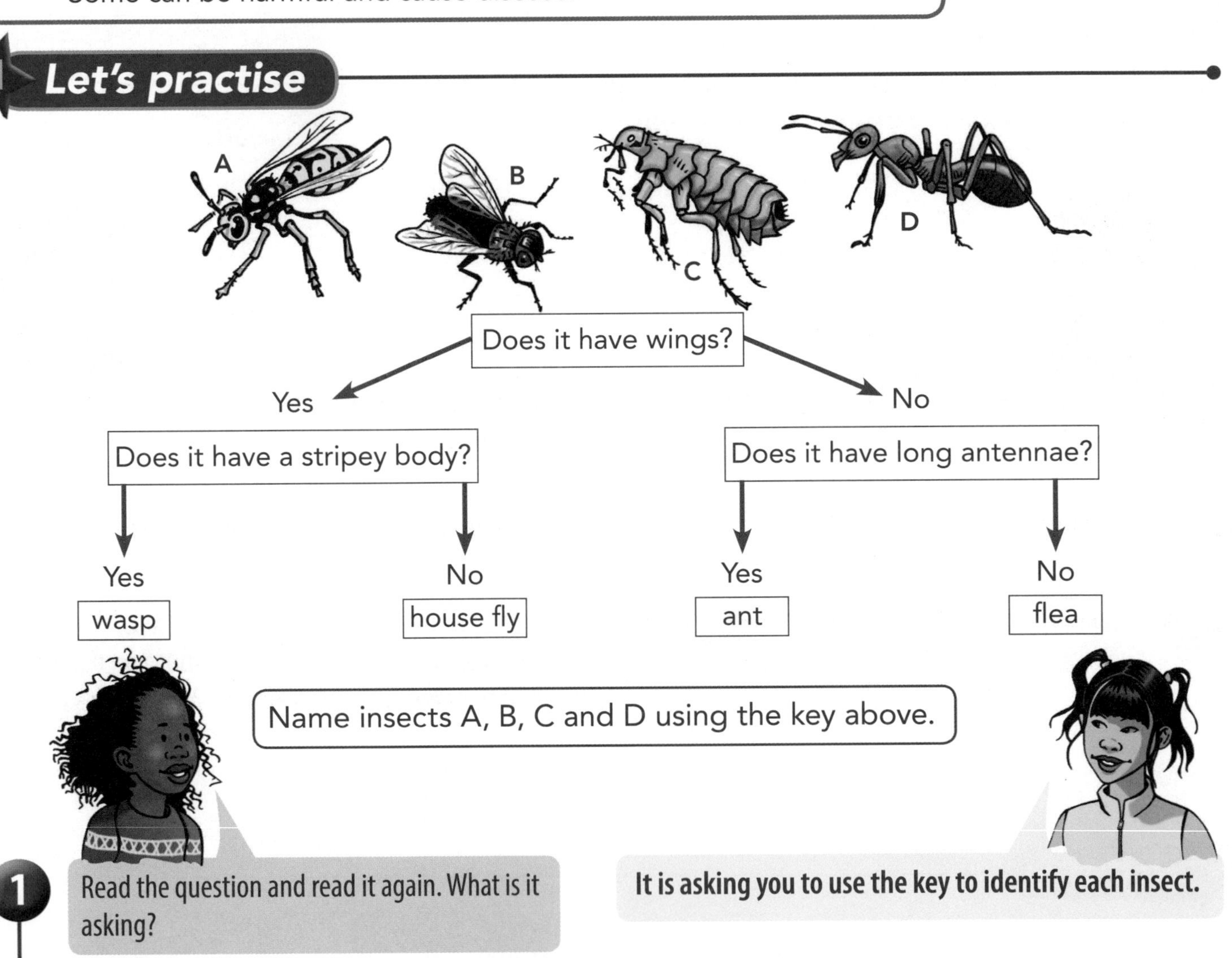

Name insects A, B, C and D using the key above.

1. Read the question and read it again. What is it asking?

   It is asking you to use the key to identify each insect.

2. Look at the diagram and work through the problem.

   Think about each insect in turn and answer each question *Yes* or *No*.

3. Check your answers.

What fits where?

What you need to know

- **Vertebrates** have internal skeletons. They can be grouped into **mammals**, **birds**, **amphibians**, **reptiles** and **fish**.
- **Invertebrates** can have an external skeleton or hard shell. Invertebrates include insects, spiders, worms and snails.
- There are flowering plants and non-flowering plants.

Let's practise

The children are talking about different groups of animals. Explain whether you agree with what each child is saying.

Reptiles give birth to live young.

Amy

Amphibians spend some of their lives in water and some on dry land.

Sanjay

Amphibians have a dry, scaly skin.

Ben

1. Read the question and read it again. What is it asking?
   It is asking you to say which information is correct and which is not.
2. Remember the key facts.
   What do you know about amphibians and reptiles? What are the differences?
3. Work through the problem.
   Read through each statement again and write whether you think it is correct. Explain your answers.
4. Check your answers. You should have three separate answers.

Try this

1 Write the names of **two** animals that could go in each of the following groups:

mammals __________ __________

birds __________ __________

reptiles __________ __________

amphibians __________ __________

fish __________ __________

2 Why did the scientist Carl Linnaeus want to give everything a scientific name?

____

____

3 Draw a branching key to identify each of these leaves:

Top tip

- You can use the same question more than once if it is in a different part of the key.

Inheritance

To achieve 100 you need to:
★ show how living things are similar but not identical to their parents
★ give examples of features that are not **inherited**.

What you need to know

- Parents can pass down their features to their **offspring**. These are known as **inherited** characteristics.
- Some features are not inherited (e.g. exercising to make big muscles or changing hair colour).

Let's practise

These kittens are from the same litter and have the same parents. They are similar but not identical.

a) Which of their features do you think may have been passed on from their parents?

b) One of the kittens is very small. What might make a difference to the size of the kitten?

1	Read the question and read it again. What is it asking?	**It is asking you about inherited features and non-inherited features.**
2	Look at the picture and relate it to the questions.	**Which different features of the kittens could have been passed on from their parents?**
3	Work through the problem.	**a) Include as many possible inherited features as you can in your answer.**
4	Check your answer.	**b) Think what factors could have affected the growth of the smallest kitten.**

Try this

1 Name **one** characteristic that you may have inherited from your parents.

____

Adaptation and change

To achieve 100 you need to:
★ identify how animals and plants are **adapted** to suit their environment
★ describe how adaptation may lead to **evolution**
★ recognise that environments can change and that this can sometimes pose dangers to plants and animals that live there.

What you need to know

- Plants and animals have special features to help them survive in particular **habitats**. This is called **adaptation**.
- **Evolution** is when living things change over a long time. This can lead to new types of plants and animals.
- All plants and animals in an environment depend on each other.
- Human activity endangers some plants and animals.

Let's practise

Emperor penguins live in Antarctica where the climate is very harsh, with extemely low temperatures and fierce storms. They eat fish, which they catch in the surrounding seas. They breed and raise their chicks on the ice. Explain how their features help them to survive in their habitat.

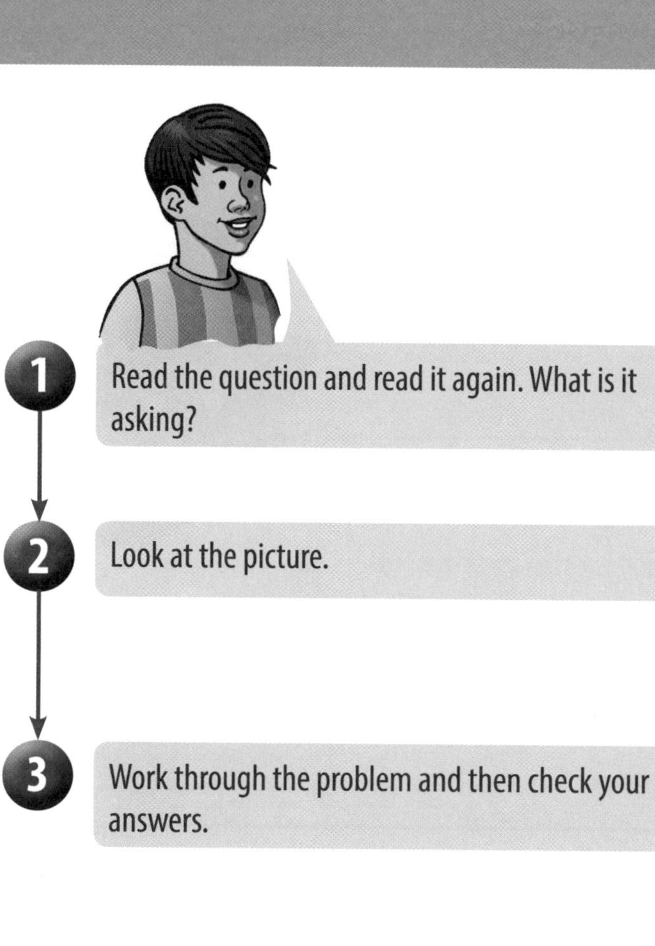

1. Read the question and read it again. What is it asking?

   It is asking you to use the information about the penguins' features to explain how they help the penguins to survive.

2. Look at the picture.

   Which feature relates to which piece of information? What helps them to keep warm? What helps them to catch their food? What will help them move around on the ice?

3. Work through the problem and then check your answers.

   Do your answers match the information you were given?

Try this

1 Emperor penguins are a threatened species because of global warming. Suggest what changes may be happening to their environment.

____

____

2 Which features of emperor penguins would be less suited to this different habitat?

____

____

3 Name a different animal that is endangered and explain why.

____

____

Investigating plants

To achieve 100 you need to:

★ identify different parts of flowering plants and describe their function
★ explore what plants need to survive.

What you need to know

- Most flowering plants have **leaves, roots, stems** and **flowers.**
- Each part of a plant has a special job to do to keep the plant alive.
- Not all plants need the same conditions to grow.

What does what?

1 Let's practise

a) Complete the labelling on the picture.
b) Describe what you would see if you put a white flower in a glass filled with blue coloured water. Explain why this would happen.

Read the question and read it again. What is it asking?

Part a) is asking you to describe the function of the different parts of the plant. Part b) is asking you to describe what would happen in an experiment and to explain why.

Look at the picture and then work through the problem.

a) Use the key facts to relate each part of the plant to its function. Are there any parts of the plant that have more than one function?
b) Have you seen this experiment done? Think about the function of the stem and try and work out what will happen.

Check your answers.

Let's find out 

2 Let's practise

I think you could cut half of the leaves off a plant and it would keep growing.

Sanjay

I think you could cut the leaves off a plant and it would keep growing.

Amy

A plant cannot live without leaves. If you cut them off it would die.

Ben

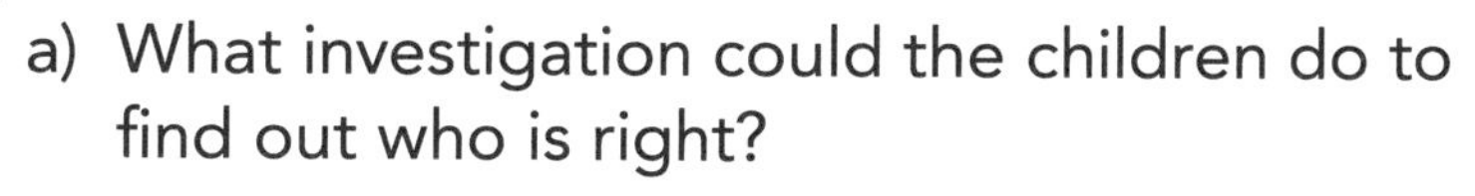

a) What investigation could the children do to find out who is right?
b) What should they keep the same during their investigation?
c) Why is it difficult to plan a fair test with plants?

1 Read the question and read it again. What is it asking?

It is asking you to plan an investigation to see the effect of leaves on a plant's growth. It must be a fair test.

2 Remember the key facts about fair testing.

You need to change one thing and keep everything else the same.

3 Work through the problem.

a) Think of an investigation that would test each of the children's ideas. You need to compare the three ideas.
b) What should they do to all of the plants to make it a fair test? Think of as many things as you can.
c) Can you be sure of how plants will behave?

4 Check your answers.

Try this

1 What do you think might happen in this investigation? Explain your answer.

____

____

Top tip

- Plants are living things and do not always behave in the same way, so use more than one plant in an experiment if you can.

The life cycle of flowering plants

To achieve 100 you need to:
★ explain how flowering plants **reproduce**.

What you need to know

- **Pollination** is when pollen moves from the male part of the flower (**stamen**) to the female part (**stigma**) for **fertilisation** to happen.
- Sometimes the stigma is in the same flower as the stamen and sometimes it is in a different flower.
- Seeds need to be **dispersed** away from the parent plant to prevent overcrowding.
- When a seed starts to grow it is called **germination**.

Let's practise

This diagram shows the different parts of a flower. Complete the labelling on the picture.

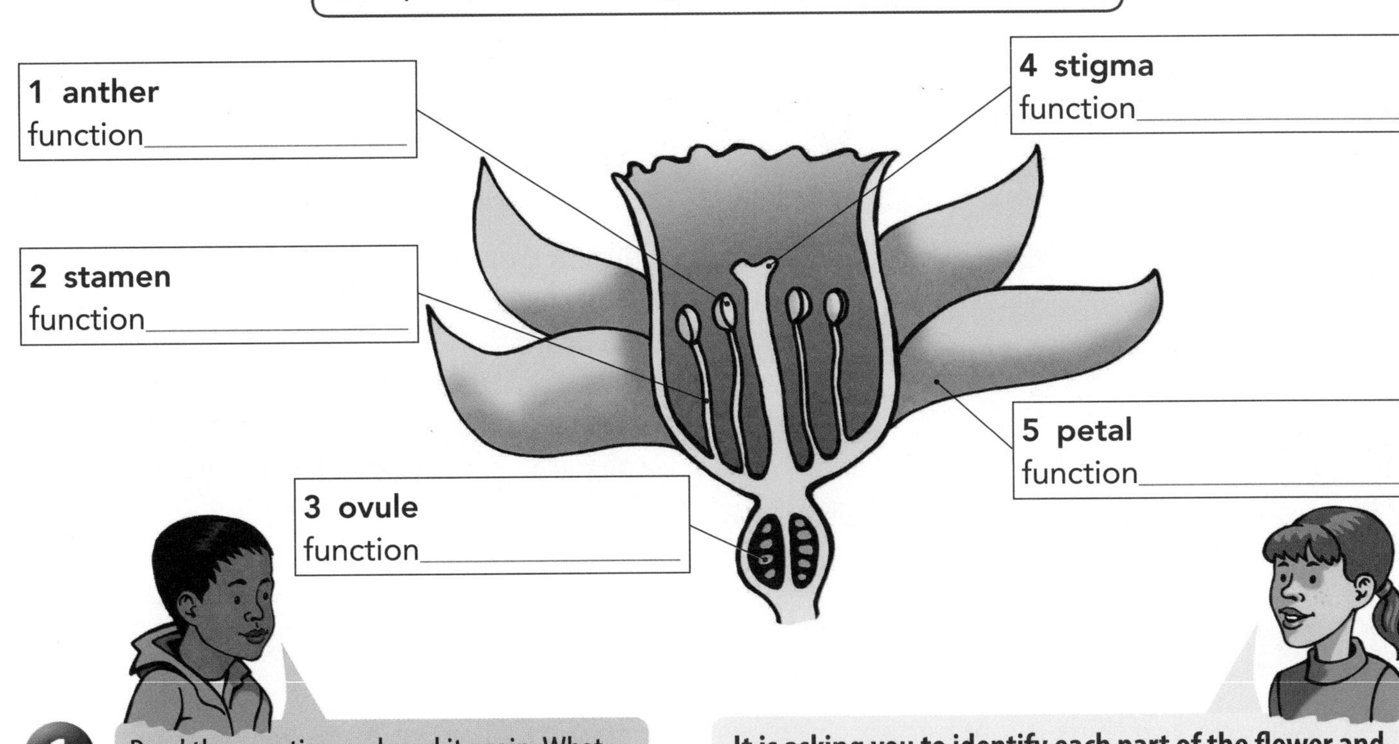

1 Read the question and read it again. What is it asking?

It is asking you to identify each part of the flower and the job it does.

2 Remember the key facts.

Start with the parts you know. Use the key facts to link each part of the flower to its job. Which parts are male and which are female?

3 Work through the problem and then check your answers.

Try this

1 Put the main stages in the life cycle of a plant in order. Write the name of each stage in the correct circle.

pollination germination seed dispersal flowering fertilisation

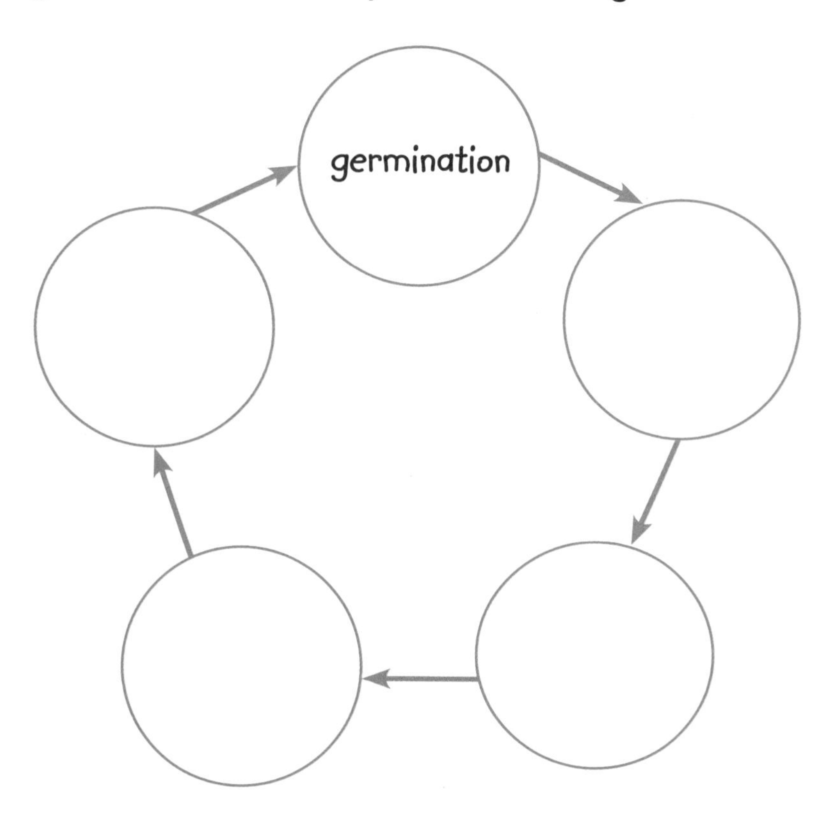

2 Explain why some flowers need insects to complete the plant's life cycle.

____

____

3 Write **two** ways that flowers can attract insects.

____

____

4 Describe **two** different ways seeds can be dispersed.

____

____

Top tip

- StaMEN is the male part of the flower and stigMA is a female part.

Rocks and fossils

To achieve 100 you need to:
- ★ compare and group rocks
- ★ describe how **fossils** are formed
- ★ describe how some living things have changed over time
- ★ describe how **soils** are made.

The rocks beneath your feet 

What you need to know

- Rocks can have different properties. They can be hard or soft, **permeable** or **impermeable**. They can have crystals, and can have layers.

1 Let's practise

Abi and Sanjay are trying to find out which rocks would make the best statues. Here are the results of some tests that they did:

Name of rock	granite	slate	marble	chalk	sandstone
Is it permeable?	✗	✗	✗	✔	✔
Is it hard?	✔	✗	✔	✗	✗

a) Which rock would you choose for a statue? What information from the table did you use to help you decide?

b) How could Abi and Sanjay test rocks to see if they are permeable?

1. Read the question and read it again. What is it asking?

Part a) is asking you to choose the best rock for a statue and explain why. Part b) is asking you to write an experiment to test if rocks are permeable.

2. Look at the table and remember the key facts.

What is the table showing you about each type of rock? What does *permeable* mean?

3. Work through the problem. Check your answer.

a) Which properties of rock are important for a statue to exist for a long time?

b) Have you done a test like this? What simple test could you do to see if a rock is permeable?

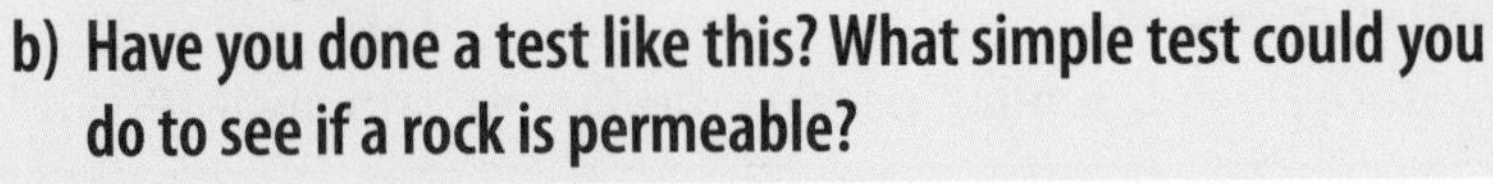

Fossils

What you need to know

- **Fossils** are the remains or impressions of living things that have turned to stone.
- Fossils provide information about living things that inhabited the Earth millions of years ago.
- Fossils can be formed in different ways. Most are made when the plant or animal is buried by mud or sand, frozen in ice or dried out in deserts or caves. Some are made from impressions left by animals (e.g. footprints).

2 Let's practise

Draw a line to match each fossil picture to the creature it came from.

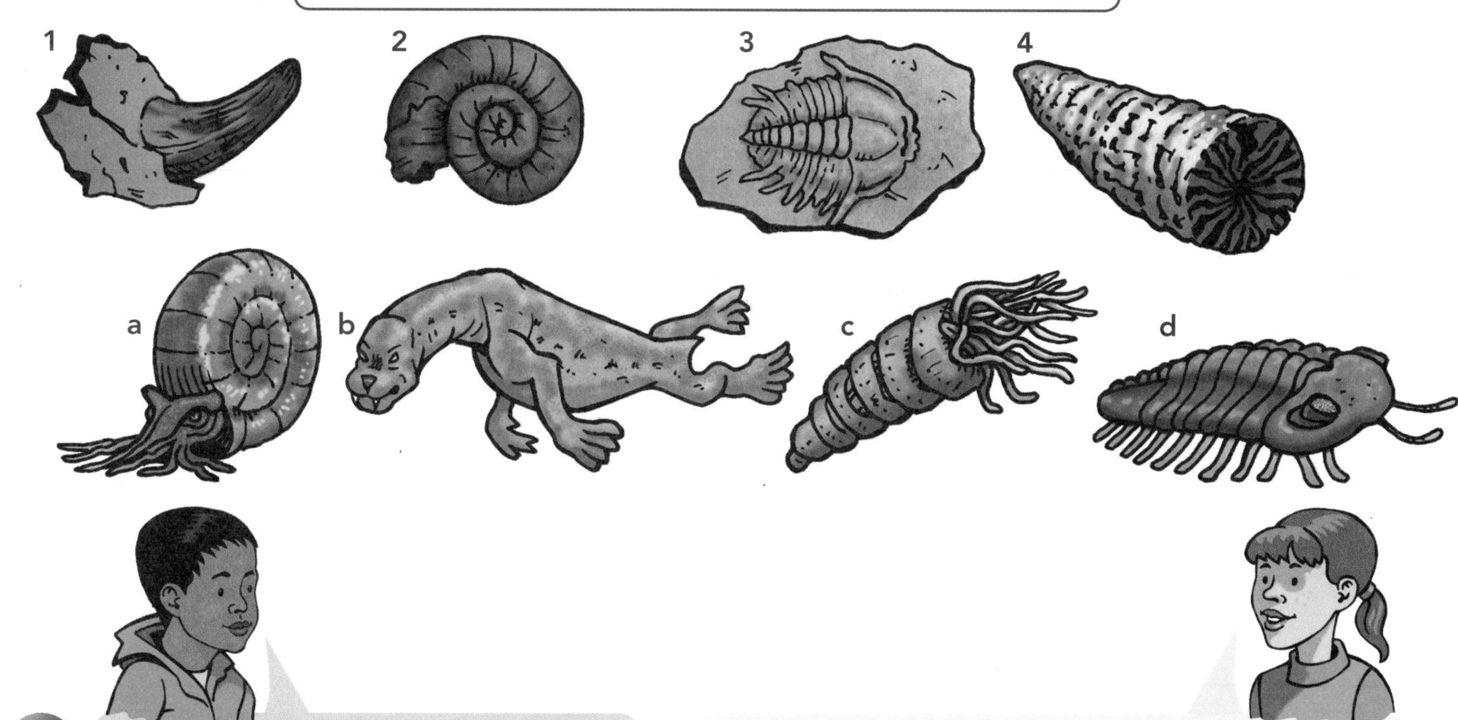

1. Read the question and read it again. What is it asking?

   It is asking you to match the fossils to extinct creatures.

2. Look at the pictures.

   Decide which features of the creatures could have been turned into the fossils shown. Remember that sometimes only part of a creature becomes a fossil.

3. Work through the problem. Then, check your answer.

   Have you matched all the pictures?

Where the worms live

What you need to know

- **Soil** is made from broken-down rocks and dead plants and animals.
- Soil can also contain living things, nutrients, air and water.
- Some soils hold onto water (e.g. clay) and some drain well (e.g. sandy soil).

3 Let's practise

This drawing shows what happens if you mix soil with water and then let it settle. Label the diagram.

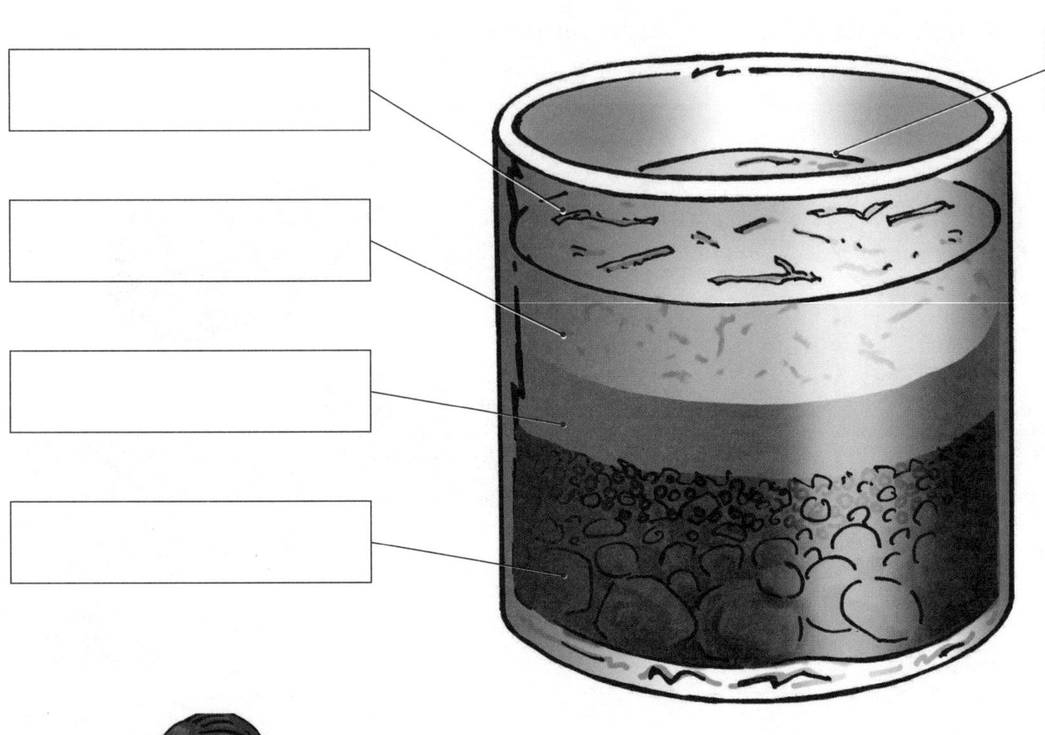

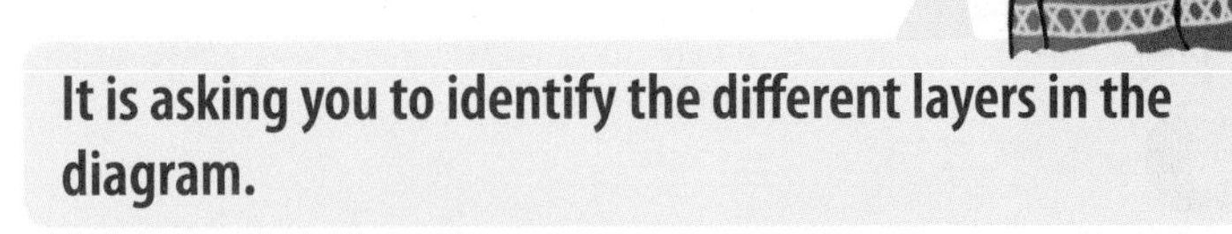

1. Read the question and read it again. What is it asking?

   It is asking you to identify the different layers in the diagram.

2. Look at the picture and remember the key facts.

   What is soil made of? Why are there different layers? How does the size of the particles differ as you move down? What is likely to be floating on the surface?

3. Work through the problem. Check your answers.

   Is your diagram labelled correctly?

1 Write a step-by-step description of how a fossil can be formed.

____

____

____

2 Explain what kind of information fossils can give us about animals that lived millions of years ago.

____

____

3 What kind of information are fossils **not** able to give us?

____

4 This soil is from a wood. Explain why the soil in a wood contains lots of organic material.

____

____

5 Describe some of the ways that the soil from the wood could be different from the soil the children tested in the *Let's practise* question.

____

____

6 Which soil do you think would be best for worms?

____

____

Solids, liquids and gases

To achieve 100 you need to:
★ decide if materials are **solids, liquids** or **gases**
★ describe how some materials change **state** when they are heated or cooled
★ describe how a change of state is a **reversible** change.

How can you tell?

What you need to know

- Materials can be grouped into **solids**, **liquids** and **gases**, depending on their properties.
- Some materials can be tricky to group.

1 Let's practise

You can hold a solid in your hand and it stays the same shape.

You can always tell when something is a liquid because you can pour it.

Gases are very light and always rise to the top of any container you put them in.

Ben

The children are grouping some materials into solids, liquids and gases.

a) Explain whether you agree with each child and give your reason why.

b) Why might some children think salt is a liquid? How can you prove it isn't?

1. Read the question and read it again. What is it asking?

   Part a) is asking you to explain why each child's statement is correct or incorrect. Part b) is asking you to discuss the properties of salt.

2. Remember the key facts.

   What are the properties of solids, liquids and gases? Do any of them have the same properties as each other?

3. Work through the problem. Then, check your answers.

   a) Read through each statement again and write whether you think it is correct. Explain your answer.
   b) Think about which properties of solids and liquids salt has.

Top tip

- Some solids, like sand, are made of particles and can be poured.

Changing state

What you need to know

- Materials change **state** when they are heated or cooled.
- Heating can change a solid to a liquid (**melt**) and a liquid to a gas (**evaporate**).
- Cooling can change a gas to a liquid (**condense**) and a liquid to a solid (**freeze**/solidify).

2 Let's practise

Write the name of the change of state in each of the four boxes. Write the temperature at which these changes happen in box 1 and the temperature at which water boils in box 2.

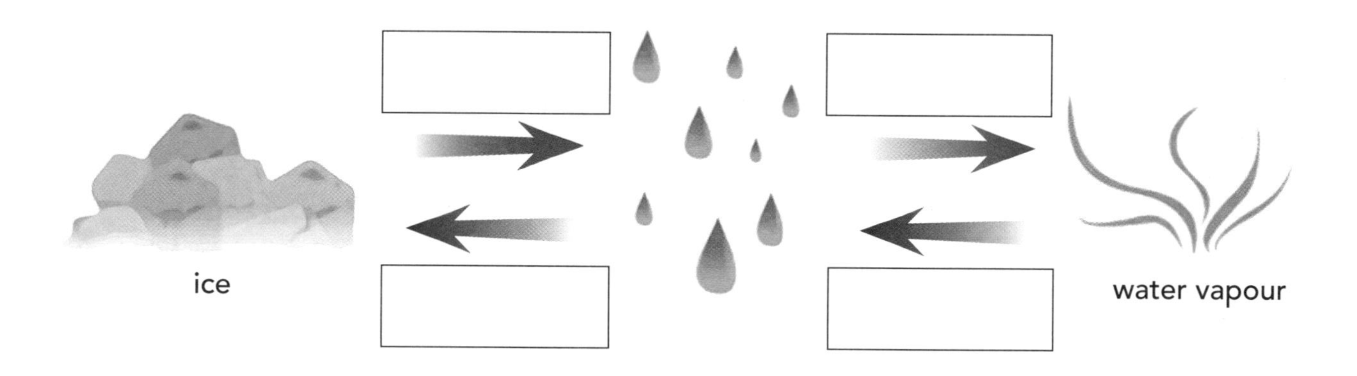

1 ________ °C

2 Boils at ________ °C

1 Read the question and read it again. What is it asking?	It is asking you to name the changes of state when water is heated or cooled.
2 Look at the picture and think of real-life examples. Remember the key facts.	What happens at each change of state? How hot or cold is it when these changes happen?
3 Work through the problem. Then, check your answers.	Write the names using the correct scientific vocabulary. Add the temperatures.

Top tip

- The temperature at which different materials change state varies.

Can you get it back? 

What you need to know

- Some changes are **reversible** so we can get back what we started with (e.g. changes of state, **dissolving**, **sieving** and **filtering**).

Let's practise

Emily takes an ice lolly out of the freezer and puts it on the table. Her friend arrives and they go out, forgetting about the ice lolly.

a) What had happened to the ice lolly when Emily returned two hours later? Explain why this happened.

b) Can Emily get her ice lolly back to a solid? Explain your answer.

1	Read the question and read it again. What is it asking?	**Part a) is asking you to identify and explain a change of state. Part b) is asking you to describe how to reverse the change.**
2	Remember the key facts.	**What makes something change state? What are the correct words to describe the changes?**
3	Work through the problem.	**a) Think about this as a real-life example. What happens to an ice lolly if you do not eat it straight away? Use the correct word to describe the change. What causes the change?**
4	Check your answers.	**b) What change of state needs to happen? Where would it be possible for this change to happen?**

Try this

1 Why do windows steam up when it is cold outside?

____

2 Describe how you could change the state of **two** different foods and then get each food back again.

____

____

The water cycle

To achieve 100 you need to:
★ describe the part played by **evaporation** and **condensation** in the **water cycle**
★ explain how temperature can affect the rate of evaporation.

Let it rain

What you need to know

- The amount of water on Earth always stays the same. It keeps moving and is recycled over and over again.
- The Sun, evaporation and condensation play important parts in the water cycle.

1 Let's practise

The diagram shows the water cycle. Fill in each box to describe what is happening at each stage.

1. Read the question and read it again. What is it asking?

   It is asking you to describe what is happening at four different stages in the water cycle.

2. Look at the diagram. Remember the key facts and the correct scientific words.

   Think carefully about what is happening at each stage. At which stages does water change state?

3. Work through the problem. Then, check your answers.

   a) What is the name of the process where water changes to a gas?
   b) What is the name of the process where water changes from a gas to a liquid?
   c) How does water get back to Earth?
   d) How does water get back into the sea?

Evaporation

What you need to know

- Temperature affects how quickly water evaporates.
- Evaporation can happen at low temperatures.

2 Let's practise

The class investigates if temperature affects how quickly water evaporates. They put some dishes of water in different places in the classroom. They time how long it takes for all the water to evaporate and record the results in this table.

Place	Time taken for water to evaporate (in hours)
On heater	1.5
On windowsill	2.5
On a desk	4
On shelf by the door	5

Use the table to draw a bar graph. Label the axes. Write a title.

a) Which is the coldest place? How do the results prove this?

b) What conclusion could the class draw from their results?

c) Write **two** things that the children would have needed to do to make it a fair test.

1. Read the question and read it again. What is it asking?
   It is asking you to draw a bar graph and write about the investigation that the class does.

2. Remember the key facts.
   What happens to the water in the dishes? How does temperature affect the speed at which this happens?

3. Draw the bar graph.
   Work out the scale for the side axis (you need to go from 0 to 5 hours). What label do you need to put on this axis? What label do you need on the bottom axis? How many bars do you need? Label each bar. The title should explain exactly what the graph shows.

4. Work through the questions.
   a) Will the longest or the shortest bar represent the coldest place? What might make this place the coldest?
   b) What is the class trying to find out? Link the temperature of each place to how long it takes the water to evaporate.
   c) What would you keep the same and what would you change to make it a fair test?

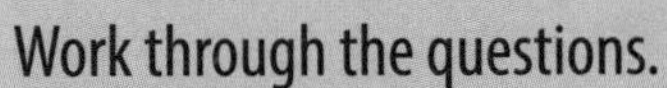

5. Check your answers.

Try this

1 Describe how rain that has fallen onto the land could get back into the sea.

2 Water does not always return to the Earth as a liquid. Name a different form it could take.

3 Where could you put your damp coat to dry quickly? Use scientific words to explain why you have chosen that place.

4 Explain why puddles dry up more quickly on a sunny day than on a cloudy day.

Top tips

- Evaporating and condensing are the reverse of each other.
- The temperature of the air affects how water evaporates or condenses.

Reversible changes

To achieve 100 you need to:
★ describe how a **solution** is made
★ decide when to use **filtering**, **sieving** and **evaporating**
★ explain how **mixtures** can be separated
★ explain why **dissolving** and mixing are **reversible** changes.

What you need to know

- If a change is reversible, you can get the original materials back.
- When a solid **dissolves** in water (or other liquid), it makes a **solution**. It is possible to get the material back by evaporating the water (or other liquid).
- Dissolving can be speeded up (e.g. by using hot water).
- If a solid does not dissolve, you can get it back by filtering or sieving.

Let's practise

The class tests some materials to see if they dissolve. Ben and Asher test salt, flour, sugar and sand by mixing each material in a different cup of water and leaving it for five minutes. They record their results as a table.

Name of material	Observation of what had happened after 5 minutes	Had it dissolved?
Salt		
Flour		
Sugar		
Sand		

a) Complete the table of their results.
b) Which materials could you get back **only** through evaporation?
c) Describe a different way that you could get the other materials back.
d) Describe another investigation the children could do to find out more about dissolving.

1	Read the question and read it again. What is it asking?	**It is asking you to decide what each material will look like after it has been mixed with water and whether it has dissolved.**
2	Look at the table.	**Use the headings to help you decide what to put in each box.**
3	Remember the key facts.	**How do you get a material back once it has disssolved?**
4	Work through the problem.	**a) Describe what both the water and the material look like after being mixed and left to stand for five minutes. Decide which of the materials will dissolve.** **b) You could get all of the materials back by evaporating the water but some of them can be got back using other ways.** **c) What different method can you use to separate materials from water if they have not dissolved?** **d) Choose one material that dissolves. Think of a question: What would happen if . . .? What investigation could you do to find the answer?**
5	Check your answers. Is your table filled in correctly?	

Try this

1 Apart from using hot water, how can you make solids dissolve more quickly?

____

____

2 You have some soil in a bucket that you are going to put into flower pots and plant seedlings in. What piece of equipment could help you to remove all the stones from the soil?

____

____

3 Uncle George says that sugar melts when it's added to tea, but you think this doesn't sound quite right. Explain what happens to it, and why.

____

____

Top tip

- Some materials can be returned to their original state in more than one way.

Irreversible changes

To achieve 100 you need to:
★ know that some changes are **irreversible** (non-reversible)
★ know that irreversible changes produce new materials.

What you need to know

- If a change is **irreversible**, you cannot get the original materials back.
- New materials are formed during an irreversible change.
- Irreversible changes can be caused by burning, heating and mixing.

Let's practise

Meena investigates what changes take place when she mixes some materials together. She puts 50 ml of water into a container and adds some plaster of Paris. She notices that the mixture gets hot and then forms a solid lump.

a) Why do you you think the mixture gets hot?
b) When the water is added to the plaster of Paris, what kind of change happens? How do you know?
c) Meena crushes some of the solid lump and mixes it with water. What do you think will happen?
d) Why do you think she does this?

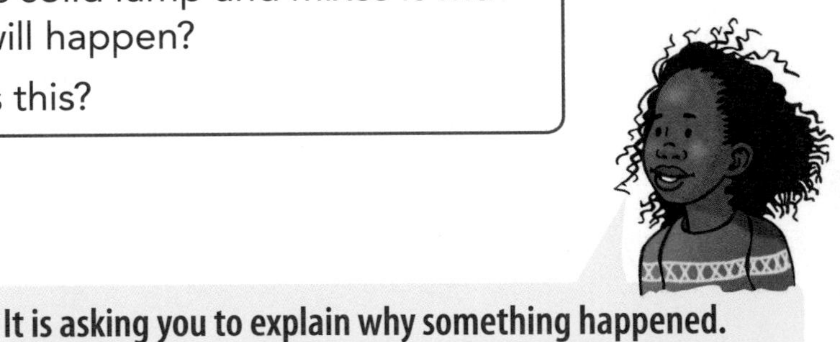

1. Read the question and read it again. What is it asking?

   It is asking you to explain why something happened. Then you have to apply what you know about irreversible changes.

2. Remember the key facts.

   What kind of changes are there? Which kind of change is this? How do you know?

3. Work through the problem.

   a) What do you think giving off heat shows?
   b) Your answer to part a) will help you to decide what kind of change has happened.
   c) Do you think it will dissolve?
   d) What was Meena investigating?

4. Check your answers.

Try this

1 Explain how an irreversible change is different from a reversible change.

2 Name **two** materials that change irreversibly when they are heated (clue: think food):

3 a) When baking powder is mixed with vinegar it fizzes. What is happening?

b) What kind of change is this?

4 Burning a piece of wood is an irreversible change. Explain why.

Top tip

- A change is irreversable (**non-reversible**) if we can't get back the materials we started with.

Properties and uses of materials 

To achieve 100 you need to:
★ compare and group together everyday materials using different **properties**
★ link properties of materials to their use
★ describe how tests are used to find out about the properties of materials.

What you need to know

- Materials can be grouped in different ways:
  - where they originate from (e.g. natural or man-made)
  - using different properties (e.g. strong, tough, absorbent).
- Materials are chosen for particular uses because of their properties.
- Some properties are more important than others.

Let's practise

The children are designing and making swimming bags. First they need to choose a suitable fabric to use.

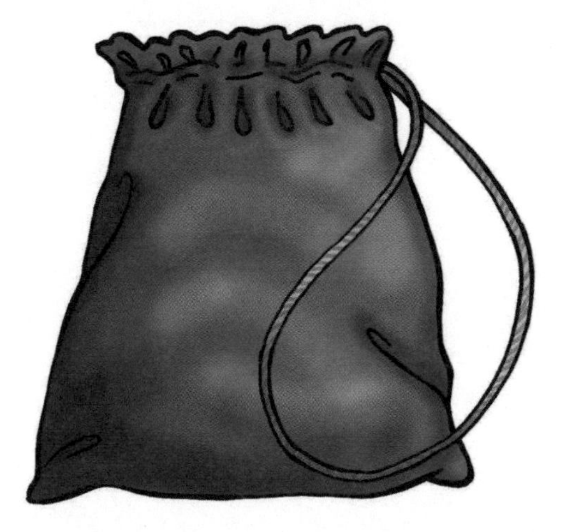

a) The teacher gives them a choice of five fabrics to use. They are going to find out which fabrics are waterproof. Describe how they could investigate this.

b) What should they keep the same to make their investigation fair?

c) What other property of the fabric could they investigate before choosing which one to use for their swimming bags?

1 Read the question and read it again. What is it asking?

It is asking you to think about the properties needed for a swimming bag. You need to plan how the children could test different fabrics and then decide on a different property that could be investigated.

2 Work through the problem.

a) What equipment will they need? How will they test the fabrics? What will they measure or observe? How will they record their results?
b) Think of as many variables as you can.
c) Choose a different property. Make sure it is a property that the children could investigate in the classroom.

3 Check your answers.

Try this

1 a) Spoons are often made of metal. Name **two** properties of metal that make it a good choice of material.

____

b) Name **one** property of metal that is not suitable for a spoon.

____

c) Write **two** other materials that could be used to make a spoon and explain why each would be a good choice.

____

____

2 a) Finish the following sentence:
If something is a thermal conductor, it ______________________.

b) Give an example of a material that is a thermal conductor and suggest where the material is used.

____

Top tip

- It is easy to confuse properties like *tough*, *hard* and *strong*. If an object is tough, it will not break or crack but it might change shape; if it is strong, it will take a heavy load without breaking or changing shape; if it is hard, it will not scratch or change shape easily but it could break or crack.

What uses electricity?

To achieve 100 you need to:
★ identify common appliances that use **electricity**.

What you need to know

- Most of the electricity that we use comes from the mains supply or from **batteries**.
- Mains electricity can be extremely dangerous. It is important to be safe when using electricity.
- Chemicals inside batteries react to make an electric current. Some electrical appliances can run both on the mains and batteries.
- Some batteries can be recharged.

Let's practise

a) Write the names in the Venn diagram to show which appliances run on mains and/or batteries:

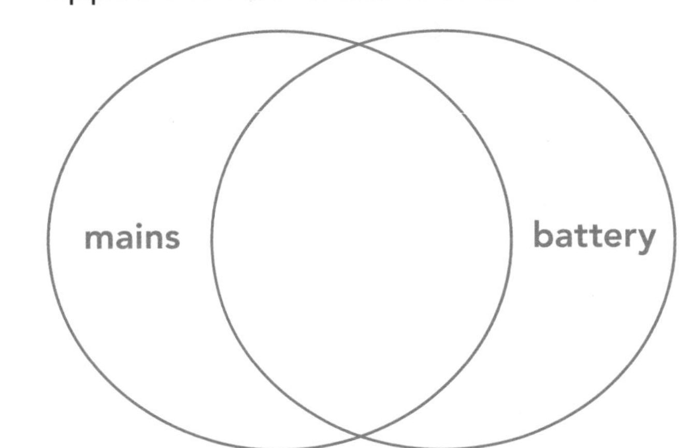

- laptop
- washing machine
- remote control
- torch
- cooker
- freezer
- radio
- watch

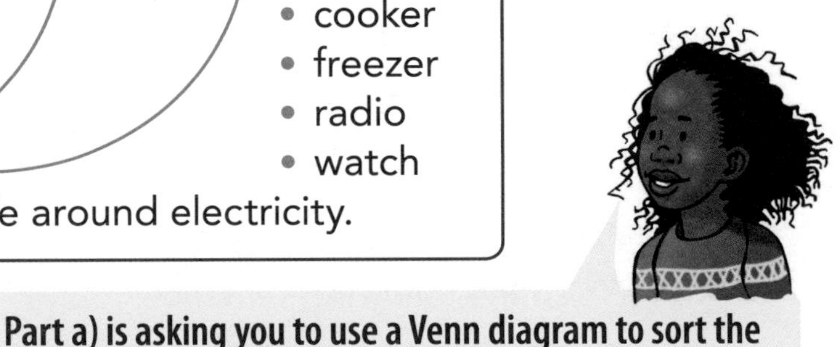

b) List **two** ways to keep safe around electricity.

1 Read the question and read it again. What is it asking?

Part a) is asking you to use a Venn diagram to sort the items. How many different groups can be shown on this Venn diagram? In part b) you need to give two ways to be safe with electricity.

2 Look at the list. Then, work through the problem.

a) Consider each item in turn. Does it have a plug? If it does have a plug, does it always have to be connected to the mains?
b) How can electricity harm us?

3 Check your answers.

a) Have you put each item in the correct place? Make sure you have identified those that use mains and batteries.
b) Have you given two ways?

Circuits and circuit diagrams

To achieve 100 you need to:

★ describe how to use **cells**, wires, bulbs, **switches** and buzzers to make simple circuits
★ decide what makes a complete circuit
★ use symbols to draw a **circuit diagram**.

Will it light?

What you need to know

- A circuit is a complete loop through which electricity can flow and make components work.
- There must be a **cell** or battery in a circuit.
- Some schools use packs that act like a battery but plug into the mains.

1 Let's practise

I think that when I switch the circuit on my bulb will light because I have joined everything in a circle.

Ben

I think that when I switch the circuit on my bulb will light because I have joined everything together.

Mia

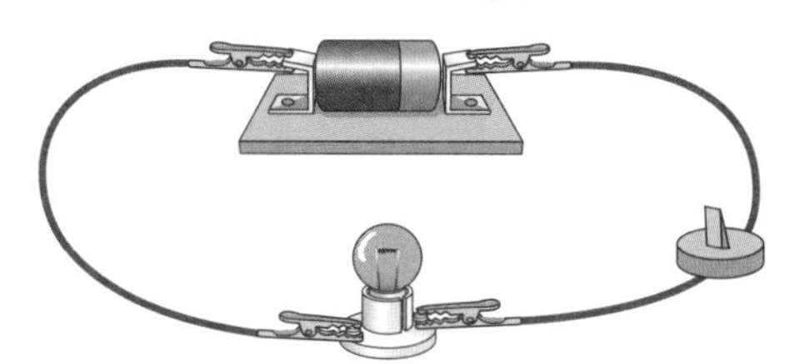

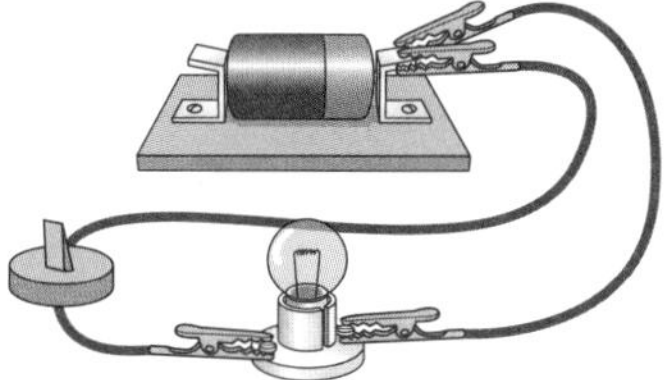

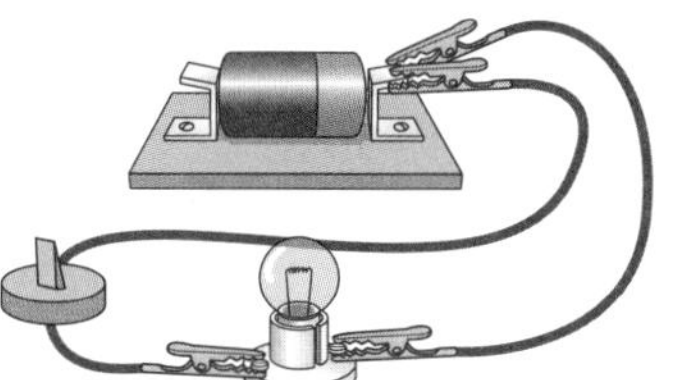

Ben and Mia are trying to make a bulb light. They each have a cell, two wires, a bulb, a switch and a bulb holder.

Do you agree with Ben and Mia? Explain your answer.

1 Read the question and read it again. What is it asking?	**It is asking you to explain why each bulb will light or not light.**
2 Look at the pictures and remember the key facts.	**Use your finger to follow where the electricity would flow. It needs to flow from the cell to the bulb and then back to the cell for the bulb to light.**
3 Work through the problem.	**Would each bulb light? You need to think about what each child says and explain whether you agree or not.**
4 Check your answers.	

Symbols and how to use them

What you need to know

- The following components have special symbols: cell, bulb, motor, buzzer, open switch and closed switch.
- You can use a diagram to represent a circuit using straight lines and right angles.
- You can use a **circuit diagram** to make a working circuit.

2 Let's practise

a) Complete the following table by drawing the symbol **or** writing the name of the component:

Component	Symbol
	—|⊢—
bulb	
	—(M)—
buzzer	
open switch	
closed switch	

b) Draw a circuit diagram that shows a circuit with two bulbs, a cell and a closed switch.

1. Read the question and read it again. What is it asking?

   Part a) is asking you to complete the table with the symbols for circuit components. Part b) is asking you to use some of the symbols to draw a circuit diagram.

2. Remember the key facts about circuit diagrams. Then, work through the problem.

   The wires much touch each component that you draw. Only draw wire on either side of each switch.

3. Check your answers.

   a) Have you filled in all the boxes?
   b) Would the circuit you have drawn light the bulbs?

1 Here is a diagram showing a circuit with a motor and a cell.

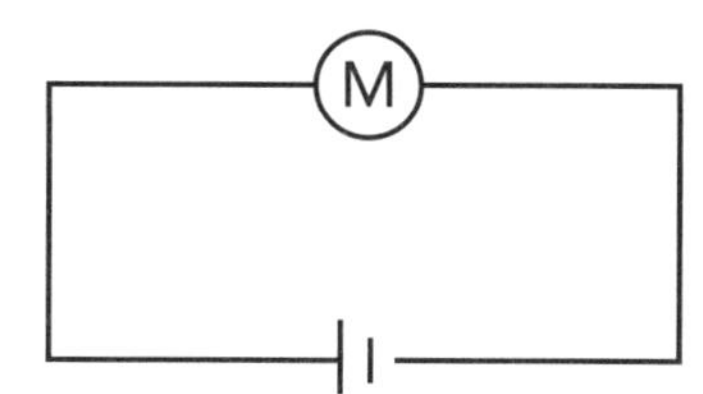

a) There is a mistake on the diagram. Circle the mistake.

b) Draw a switch in the circuit.

c) How could you make a motor turn the other way?

____

2 Draw a diagram showing a circuit with two cells, two bulbs and a switch so that the bulbs are lit.

3 A simple circuit has one cell, one bulb and a switch. Will it make a difference if the cell is turned the other way around?

____

____

Top tips

- There could be several different reasons why a bulb in a circuit will not light.
- When you draw a circuit diagram, make sure the wires are joined to both sides of each component.

Conductors and switches 

To achieve 100 you need to:
- ★ describe how **switches** are used in circuits
- ★ describe what **conductors** and **insulators** are
- ★ use correct symbols to draw a simple **circuit diagram**.

What you need to know

- A switch is a way of making or breaking a circuit.
- A **conductor** is a material that allows electricity to flow through it
- An **insulator** is a material that doesn't allow electricity to flow through it.

Let's practise

The class is making circuits. The teacher gives the children six different materials and asks them to find out which materials they could use to make a switch.

a) Describe an investigation that the children could do to find out which materials they could use for a switch.

b) Make a list of the equipment they would need.

c) Suggest one material that the teacher could give them that would make a good switch.

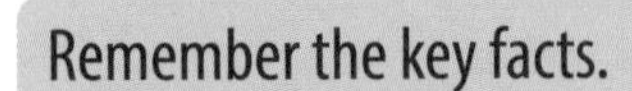

1 Read the question and read it again. What is it asking?

It is asking you to plan an investigation about making switches.

2 Remember the key facts.

What types of materials conduct electricity? What materials do you know that are insulators?

3 Work through the problem.

a) Decide exactly what you could do to find out which materials conduct electricity. Describe exactly what you would do in the investigation.

b) What equipment will you need for this investigation? Draw a diagram if it helps you to make sure you have included everything you need.

c) What materials do you know that conduct electricity?

4 Check your answers.

1

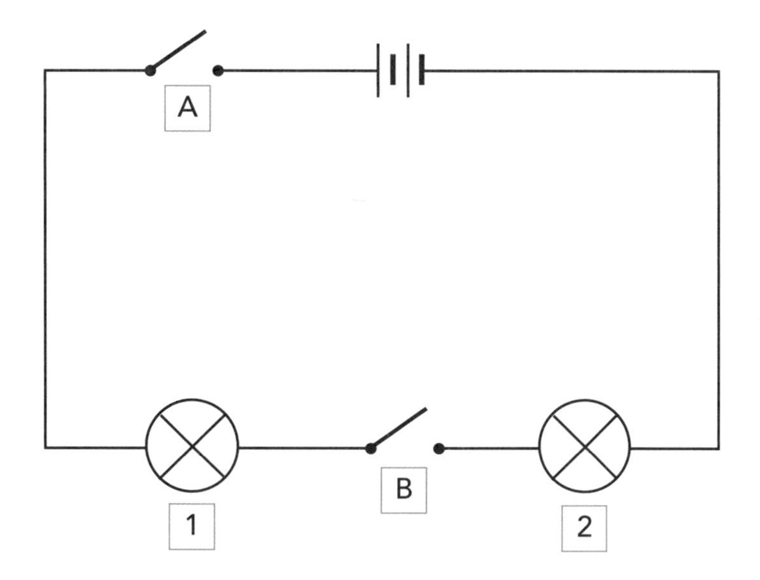

Which bulbs, if any, would light if:

a) Switch A is open and switch B is closed?

____

b) Switch A is closed and switch B is open?

____

c) Both switches are open?

____

d) Both switches are closed?

____

2 Explain why electrical wires have copper on the inside and plastic on the outside.

____

____

Top tip

- Trace the flow of electricity with your finger to see if a circuit diagram is complete.

Changing circuits   

To achieve 100 you need to:

★ describe how the number of **cells** can affect how components work in a circuit

★ use correct symbols to draw a simple **circuit diagram**.

What you need to know

- Cells can be different sizes and strengths.
- The number and types of components can make circuits work differently.

Let's practise

The children are trying to find ways of making the bulb brighter.
This is what they suggest:

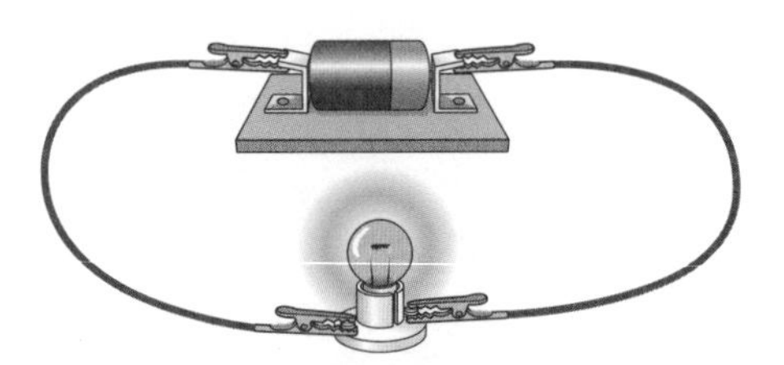

Jo

If we put two more bulbs in the circuit, we will get twice as much light.

Safi

If we put another cell in the circuit, the bulb will be brighter.

Sam

Putting six extra cells in the circuit will make the bulb much brighter.

a) Do you agree with each child? Give your reasons.

b) Draw a circuit diagram of the circuit you think would be the best one to use to make the bulb brighter.

Top tip

- Check that the total **voltage** from the cells is suitable for the components in the circuit.

1. Read the question and read it again. What is it asking?

   Part a) is asking you to comment on what each child is saying, explaining what you think of their ideas. Part b) is asking you to draw a circuit diagram of the idea you think is best.

2. Think about the question.

   Look at the diagram to help you. When you have decided which idea is best, draw a rough diagram to match the idea. This will help you to draw the circuit diagram.

3. Work through the problem.

   a) Think about what each child says separately. What would happen if you change the circuit in the way that each child suggests? b) Which symbols do you need for your circuit diagram? Use a ruler to help you draw the lines.

4. Check your answers.

   Do your answers make sense and explain what would happen in each case? Are the wires touching each component in your circuit diagram?

Try this

1 The class wants to find out if adding another cell to a circuit makes a buzzer louder. They make a circuit with a switch, a cell and a buzzer. They use a datalogger to measure how loud the buzzer is. They then add more cells and record the loudness of the buzzer. This bar chart shows their results.

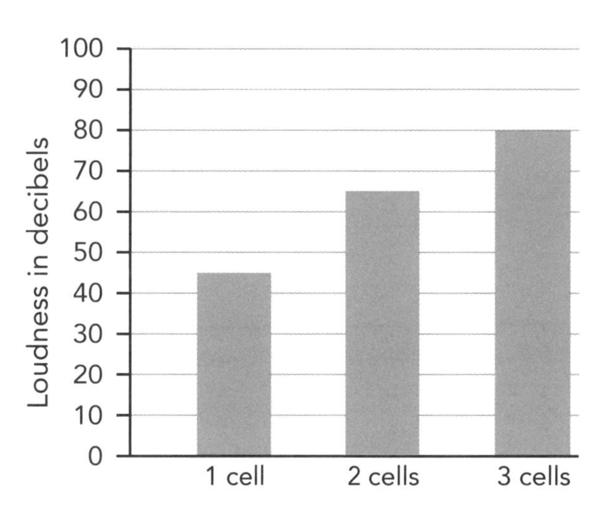

a) What conclusion could the class draw from their results?

____

____

b) Explain why it would have been a good idea for the children to have repeated their readings. Give **two** reasons.

____

____

Gravity and resistance  

To achieve 100 you need to:
★ describe **gravity** as a **non-contact force**, and describe the effects of gravity
★ describe how **air resistance, water resistance** and **friction** can slow movement.

What you need to know

- An object is attracted to the Earth because of **gravity**.
- The **weight** of an object is a measure of the force of gravity acting upon it.
- We measure force with a forcemeter and the units are **Newtons** (N).
- **Friction**, **air resistance** and **water resistance** are all stopping forces.
- Friction, air resistance and water resistance are contact forces.

Let's practise

Four children are trying to find out whose shoe has the best grip. They measure the force that it takes to drag each shoe across a surface.

a) What is the name of the force that is stopping each shoe moving?
b) Would it be a good idea for the children to repeat their readings? Explain why.
c) Will the shoe with the highest or lowest measurement of force be the one with the best grip? Explain why.

1. Read the question and read it again. What is it asking?

   It is asking you to use what you know about forces to explain the results of the investigation.

2. Look at the picture and remember the key facts.

   What do you know about measuring forces?

3. Work through the problem. Then, check your answers.

   a) Use your knowledge of forces.
   b) How accurate are their measurements likely to be? Would they get exactly the same results if they repeated them?
   c) Would a greater force be needed to move the shoe with the most grip?

Try this

1 Why don't people fall off the Earth in Australia?

____

____

____

2 Sam drops two pieces of flat paper from the same height. Explain why it takes longer for the large piece of paper to reach the ground than the small piece of paper.

____

____

____

3 Sometimes friction is a useful force and and sometimes it not useful. When would friction not be useful?

____

____

____

Mechanisms

To achieve 100 you need to:

★ describe how simple machines including **pulleys, levers, springs** and **gears** can make jobs easier by increasing the effects of a force.

What you need to know

- **Levers** and **pulleys** can turn a smaller force into a larger one (e.g. lifting a heavy weight).
- Levers have a **pivot point** or balancing point, like a seesaw.
- **Gears** can turn a shorter distance into a longer one (e.g. a hand-operated food whisk).

Let's practise

The man is using a lever to move a heavy rock.

a) Identify the **pivot point**.

b) What could the man do if he couldn't move the stone with this lever?

c) Describe how this lever works.

1. Read the question and read it again. What is it asking? — **It is asking you about how a lever works.**
2. Look at the picture and remember the key facts. — **Levers are often used to increase a force.**
3. Work through the problem. — **a) Levers always work through a pivot point. Where is it on the diagram? b) How could changing the lever help? c) What happens as a result of what the man is doing? How does the lever help?**
4. Check your answers.

Try this

I think we can lift our teacher just by using a plank of wood and a brick.

1 Describe what you think Eva's idea could be. Draw a picture to show what you mean.

2 The children are making a tree house. Describe how they could use the pulley to lift everything they need into the tree.

Top tip

- The length of a lever affects how it works.

Magnetic materials  

To achieve 100 you need to:
★ describe how **magnets** attract some materials and not others
★ compare and group materials that are **magnetic** and those that aren't.

What you need to know

- Magnetic forces can act at a distance.
- Some, but not all, metals are attracted to **magnets**. They are called magnetic materials.

Let's practise

Some materials are made from magnetic materials; some are made from non-magnetic materials.

a) Sort the objects by writing each word in the correct column.

magnetic	non-magnetic	
		steel paper clip
		aluminium can
		pencil
		steel scissors
		iron nail
		steel saucepan
		gold ring

b) Add one more object to each column.

c) Explain why some keys, when they are tested, are magnetic and some are not.

1. Read the question and read it again. What is it asking?
   It is asking you to sort the objects into a table. How many groups will you sort them into?

2. Remember the key facts.
   Which metals are magnetic and which ones are not?

3. Work through the problem.
   a) Sort the objects. Have a guess if there are any you are unsure of.
   b) Think about other objects you know that are magnetic or non-magnetic.
   c) What materials can keys be made of?

4. Check your answers.
   Have you put each item that you know in the correct place?

Try this

1 The class is trying to find out which of their magnets is the strongest.

a) Describe or draw what they could do to find out.

____

____

b) What would they need to keep the same to make it a fair test?

____

____

c) What could they measure?

____

d) Would it be a good idea to repeat their test? Explain your answer.

____

2 Write down **two** everyday uses of magnets.

____

____

Top tip

- Magnetic materials are always made of metal but not all metals are magnetic.

Magnets

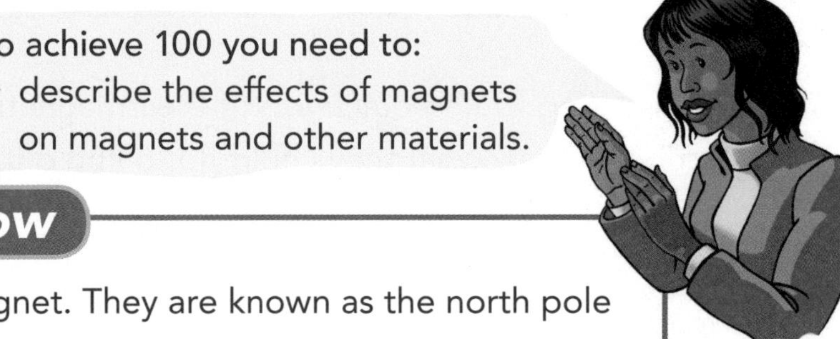

To achieve 100 you need to:
★ describe the effects of magnets on magnets and other materials.

What you need to know

- **Poles** are at the ends of the magnet. They are known as the north pole and the south pole.
- Poles that are the same (like poles) **repel** (move away from) each other.
- Poles that are different (unlike poles) **attract** (move towards) each other.

Let's practise

Meena has two bar magnets. Each magnet has a north pole and a south pole.

a) What happens if Meena moves the north end and the south end together?

b) What happens if she moves the two south ends together?

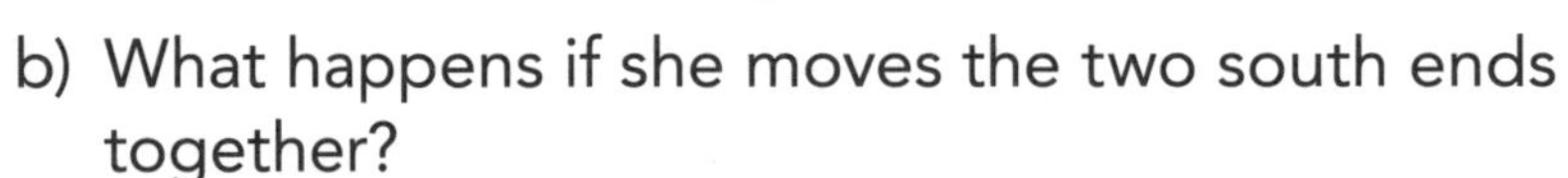

c) What happens if she moves the two north ends together?

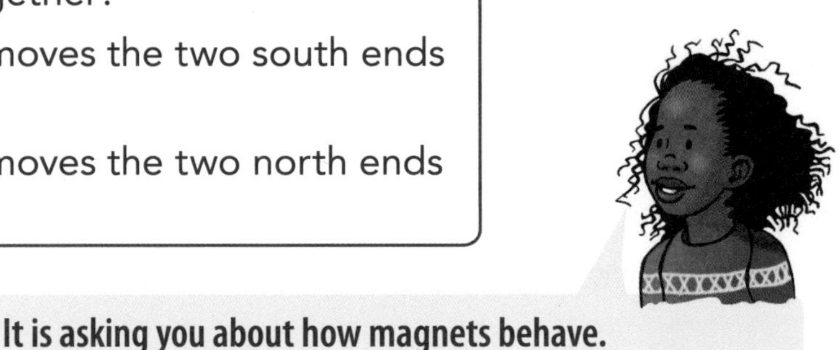

1. Read the question and read it again. What is it asking? — **It is asking you about how magnets behave.**
2. Remember the key facts. — **When do magnets attract and when do they repel?**
3. Work through the problem. — **Draw a diagram for each question if it will help you.**
4. Check your answers.

Try this

1

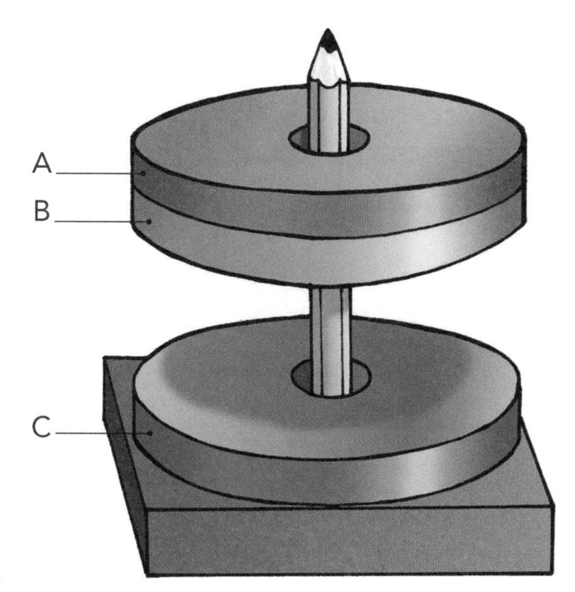

If the top face of magnet C is a north (N) pole, identify north (N) and south (S) on each magnet.

a) Why are magnets A and B touching each other?

____

b) Why is there a space between magnets B and C?

____

2 Label the north and south poles of each magnet.

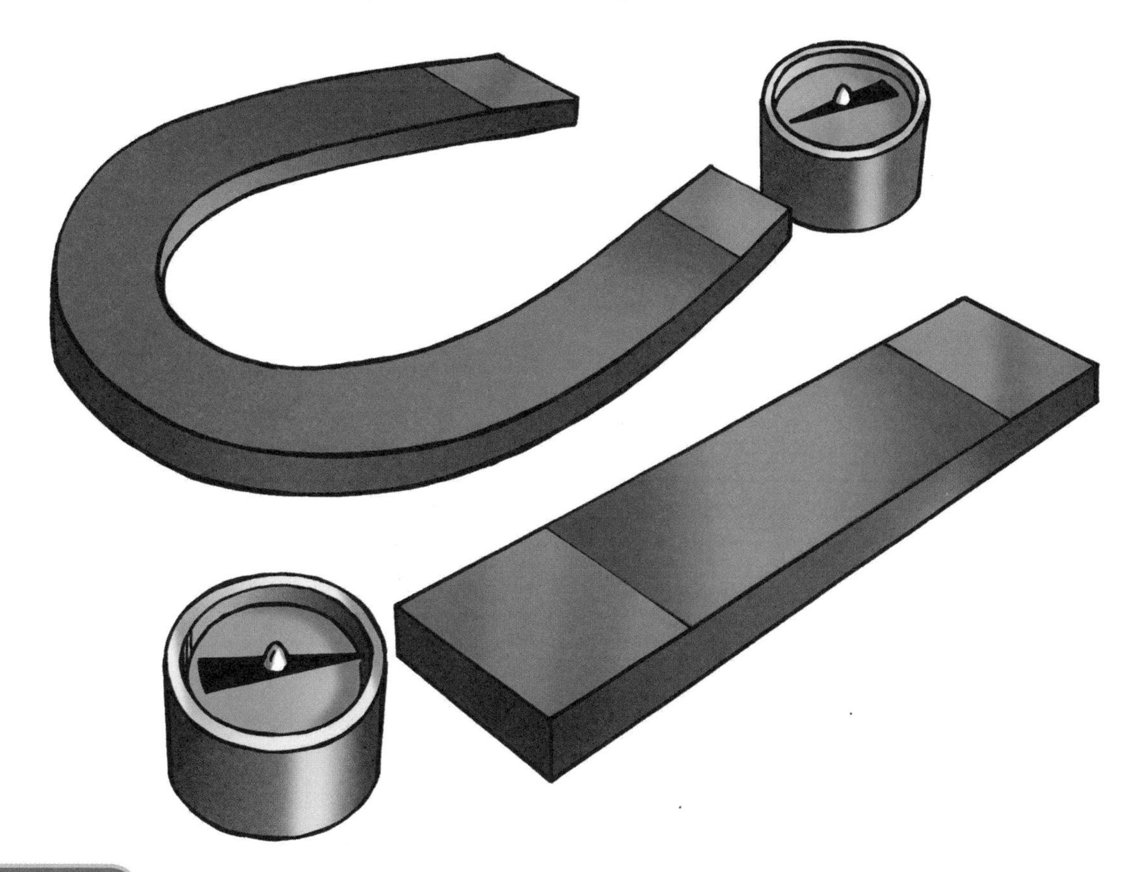

Top tips

- The magnetic force of a magnet is strongest at the poles.
- Magnetic forces are **non-contact forces**. A magnet and another magnet or magnetic material do not have to touch to attract or repel.

How we see

To achieve 100 you need to:
★ describe how we need light to see things
★ describe and draw diagrams to show how light is **reflected** from surfaces into our eyes.

What you need to know

- A light source is an object that produces its own light.
- Light travels in straight lines and it can bounce or **reflect** off **surfaces**.
- We see things because light travels from light sources to our eyes or from light sources to objects and then to our eyes.
- Light from an object gets into the eye through the **pupil**.
- The **lens** helps to make the image on the **retina** at the back of the eye clear.
- You should never look directly at the Sun.
- We can protect our eyes by wearing sunglasses in sunny weather.

Let's practise

Benedict is looking at a picture.

Draw arrows to show how the light travels to allow Benedict to see the picture.

1. Read the question and read it again. What is it asking? — It is asking you to show the way the light travels.
2. Look at the picture and remember the key facts. — You need a light source to be able to see. You see an object when light enters your eyes.
3. Work through the problem. — Start your line at the light source and finish at Benedict's eyes. How many arrows do you need to draw on the diagram?
4. Check your answers. — Use your finger to follow the arrows you have drawn. Will they allow Benedict to see the picture?

1

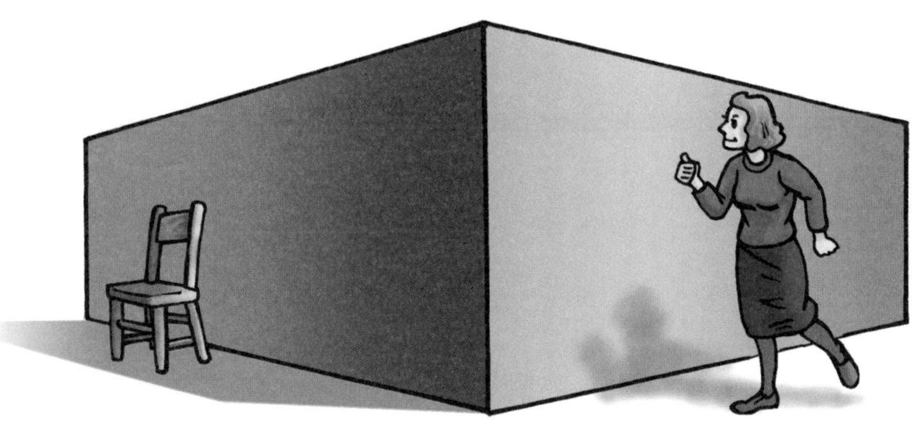

a) Why can't the person see the object round a corner?

b) Draw a mirror and arrows to show how the mirror could be positioned so that the person could see the chair.

2 Draw a diagram to show how we see the Moon.

3 Describe **two** ways to protect your eyes from the Sun.

____

____

Top tip

- Use a ruler to draw the straight lines for the arrows when showing how light travels.

Shadows   

To achieve 100 you need to:

★ describe how **shadows** are formed and how the size of shadows may change.

Making shadows

What you need to know

- **Shadows** are formed when light is blocked by a solid object.
- Different materials make different types of shadows:
  - **Opaque** materials let no light through.
  - **Translucent** and **transparent** materials let different amounts of light through.
- Shadows have the same shape as the objects that cast them because light travels in straight lines.

1 Let's practise

Aiden looks at his shadow when he is standing in the playground at lunchtime. This is the picture he draws to show what his shadow is like:

a) Explain where you think the Sun is and why.

b) Describe how Aiden's shadow is made in the playground.

c) Explain whether you think Aiden's picture of his shadow is correct.

d) What would Aiden's shadow have looked like if he had gone out into the playground just before he went home after school?

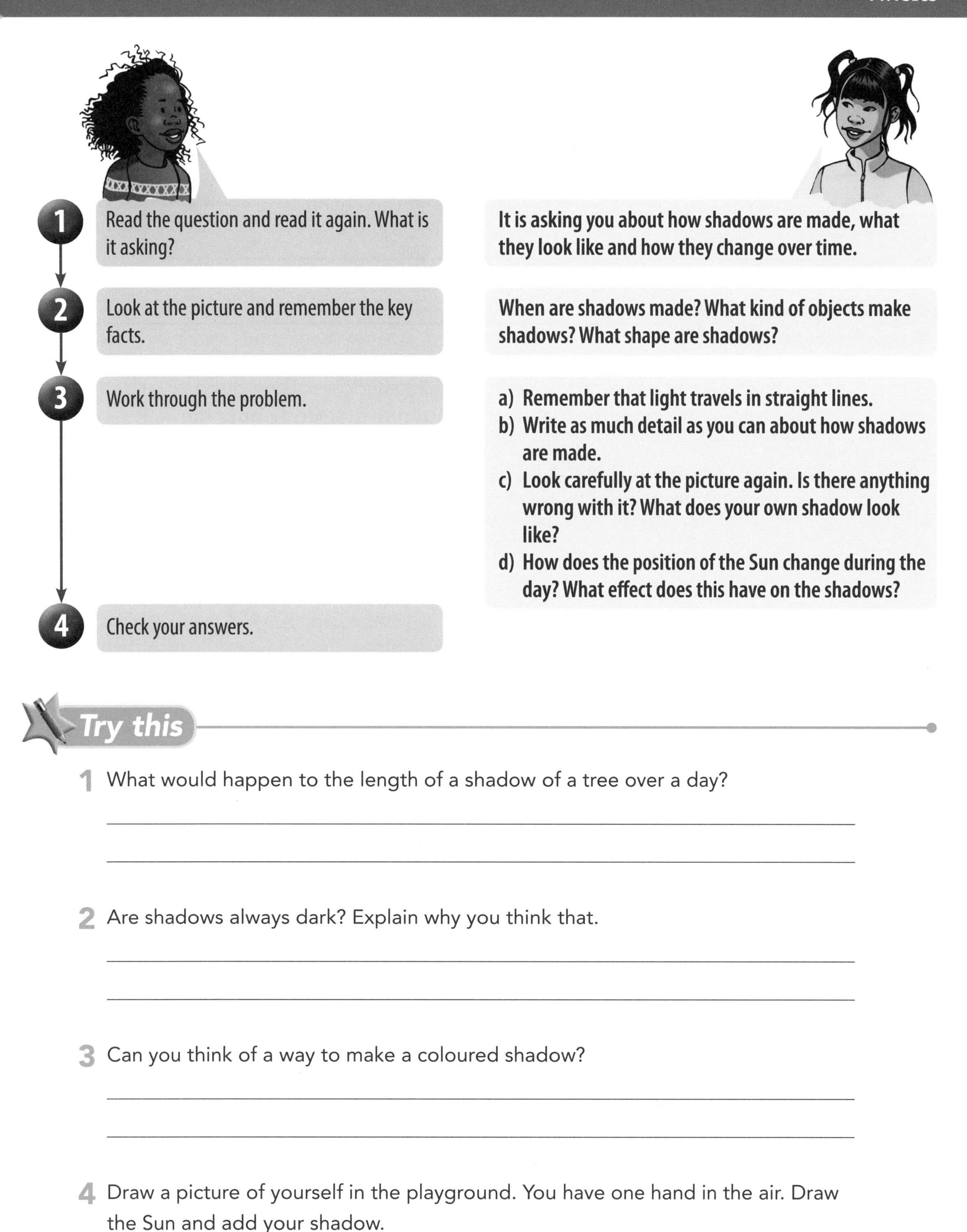

1. Read the question and read it again. What is it asking?

   It is asking you about how shadows are made, what they look like and how they change over time.

2. Look at the picture and remember the key facts.

   When are shadows made? What kind of objects make shadows? What shape are shadows?

3. Work through the problem.

   a) Remember that light travels in straight lines.
   b) Write as much detail as you can about how shadows are made.
   c) Look carefully at the picture again. Is there anything wrong with it? What does your own shadow look like?
   d) How does the position of the Sun change during the day? What effect does this have on the shadows?

4. Check your answers.

Try this

1 What would happen to the length of a shadow of a tree over a day?

____

____

2 Are shadows always dark? Explain why you think that.

____

____

3 Can you think of a way to make a coloured shadow?

____

____

4 Draw a picture of yourself in the playground. You have one hand in the air. Draw the Sun and add your shadow.

Growing shadows

What you need to know

- The size of a shadow can change if the object is moved nearer to or further from the light source.

2 Let's practise

Cody and Anna use a teddy bear to make shadows. They put a light one metre from the wall and do not move it. Cody puts the teddy bear at different distances from the wall. Anna measures the height of the shadow made by the teddy bear. They draw a graph of the measurements they take.

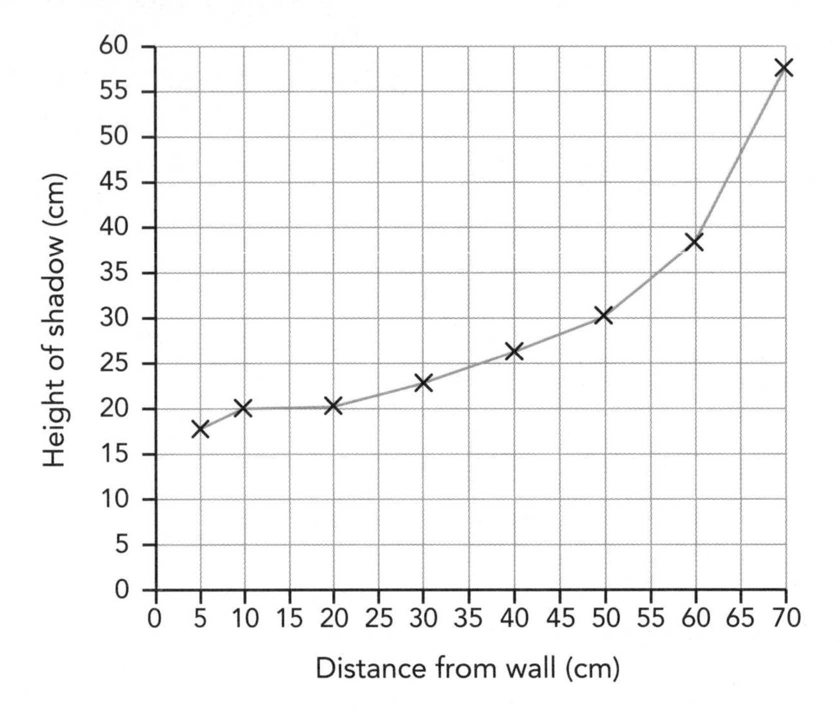

a) Write a title for the graph.

b) When the teddy bear is 50 cm from the wall, what is the height of the shadow?

c) Write a conclusion that the children could make based on their results.

d) Do you think Cody and Anna should have repeated their measurements? Explain your reasons.

1	Read the question and read it again. What is it asking?	**It is asking you to interpret a graph about a shadow investigation.**
2	Remember the key facts.	**Shadows are made when an object blocks the light.**
3	Look at the graph and work through the problem.	**a) What is the graph telling you?** **b) Find 50 cm on the bottom axis. Read up the side axis for the height of the shadow.** **c) Look carefully at the line. What is it telling you about the relationship between the distance from the wall and the height of the shadow?**
4	Check your answers.	**d) It is not always necessary to repeat readings.**

Try this

The class puts a shadow stick in the playground and measures the length of the shadows at different times in the day. They put their results in a line graph.

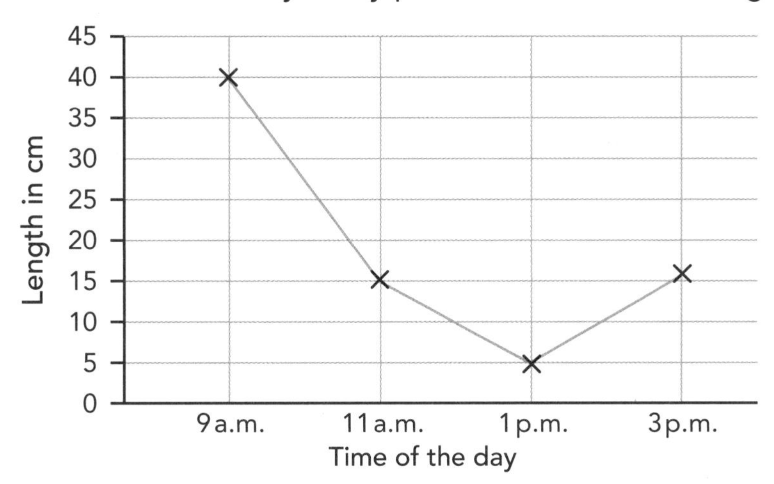

1 How long is the shadow at 11 a.m?

2 Explain why the shadow is shortest at 1 p.m.

3 How long would you expect the shadow to be at 5 p.m?

Top tip

- The positon of the light source affects the size and the position of the shadow.

Sound vibrations

To achieve 100 you need to:
★ explain that sounds are made when objects **vibrate**
★ describe the **medium** that sound travels through from a source to the ear.

What you need to know

- Sound **vibrations** travel through solids, liquids and gases to the ear.
- Sounds travel best through solids.
- Musical instruments produce sound in different ways.
- Loud noises can damage your hearing.

Let's practise

The class is learning about sound. Emma and Owain are looking at different musical instruments.

a) Do you agree with Owain and Emma? Explain your answer.

b) Explain how Owain hears the sound as Emma plays the guitar.

c) Think of a different musical instrument. Describe how the sound is made.

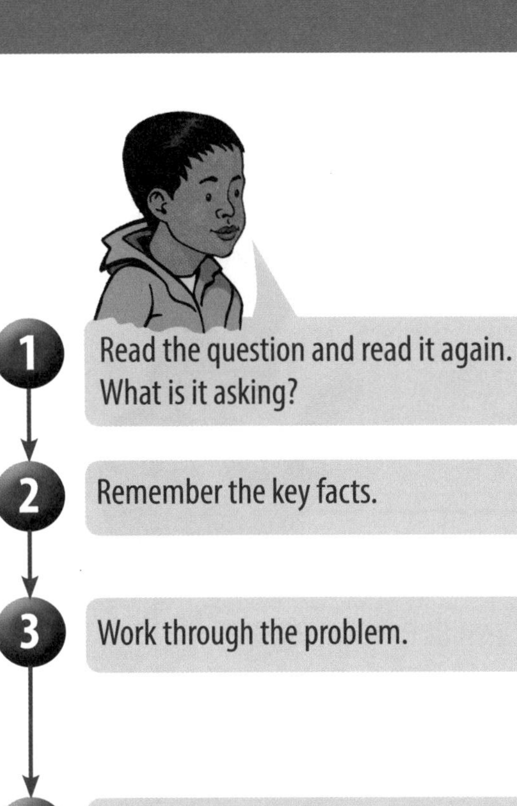

1. Read the question and read it again. What is it asking?

   It is asking you about how we hear sounds when something vibrates.

2. Remember the key facts.

   How do the vibrations reach your ears? How can you make sounds higher or lower?

3. Work through the problem.

   a) Think about what each child is saying. Do you agree with them?
   b) Explain how the sound is made and how it travels to Owain's ears.
   c) Choose an instrument. Describe how it is played and what is vibrating to make the sound.

4. Check your answers.

Try this

1 Write an example of when you hear sound travelling through:

a) a solid:

____

b) a liquid:

____

c) gas:

____

2 How could you show that the sound from a drum is made by vibrations?

____

____

3 Explain why you can still hear someone shouting in the classroom even if the door is closed.

____

____

Top tip

- Sound has to have something to travel through – this is called a **medium**.

Pitch and volume  

To achieve 100 you need to:
★ identify a link between the **pitch** of a sound and features of the object that produced it
★ identify a link between the **volume** of a sound and size of the **vibrations**.

What you need to know

- The **volume** of a sound describes how loud it is.
- Loud sounds have larger vibrations. Quiet sounds have smaller vibrations.
- The **pitch** of a sound describes how high it is.
- We can change the pitch and volume of the sounds produced by instruments.
- Sound gets quieter as the distance from the sound source increases.
- Sound travels in all directions.

Let's practise

The teacher asks the children to plan an investigation that would help them to answer the following question: 'Does sound get quieter as you travel away from the source of the sound?'

a) Describe an investigation they could do to answer the question.
b) What equipment would they need for their investigation?
c) How would they make their test fair?

1. Read the question and read it again. What is it asking?

   It is asking you to describe an investigation to answer the teacher's question.

2. Think about the question.

   What do you think the answer is? Why do you think that? What could you do to show whether your answer is right?

3. Work through the problem. Then, check your answers.

   a) Describe the investigation. What sound will you use? How will you know if the sound is quieter?
   b) Make a list of everything you will need.
   c) Do you need to keep anything the same to make the investigation fair?

Try this

1 Describe **two** ways that you could make a higher note on a violin.

2 a) What do you think vibrates when a recorder is blown?

b) How does covering the holes on a recorder change the pitch of the note?

3 Describe how you could make the sound of a drum louder.

4 How could you show that the vibrations are greater when the sound is louder?

Top tip

- The more there is to vibrate, the lower the note.

Our Solar System

To achieve 100 you need to:

★ describe the movement of the **Earth** and other planets around the **Sun**

★ describe the movement of the **Moon** around the Earth.

What you need to know

- The Sun, **Moon** and Earth are all spheres.
- All planets in our **Solar System orbit** the Sun.
- It take 365 days for the Earth to orbit the Sun.
- The Moon orbits the Earth.

Let's practise

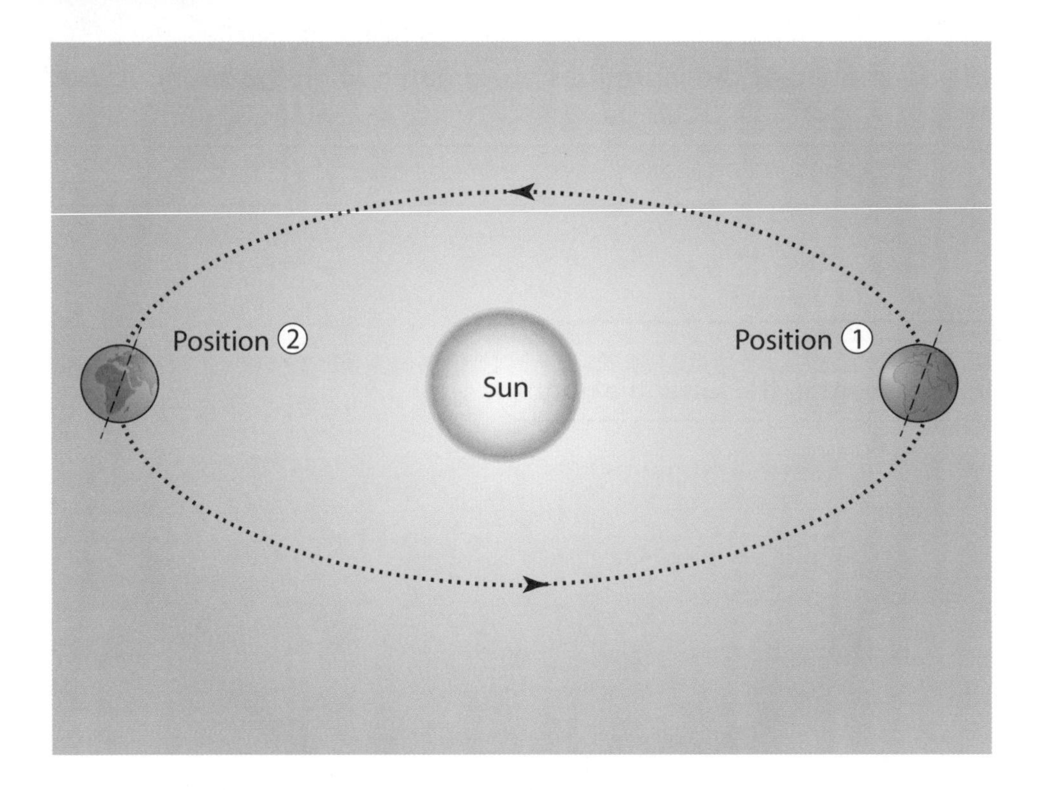

The diagram shows the Earth and the Sun.

a) The Earth is at position 1 in June. How long will it take for it to be back at this position again?

b) What month will it be when the Earth reaches position 2? What season will it be then in the UK?

1	Read the question and read it again. What is it asking?	It is asking you about the orbit of the Earth around the Sun.
2	Look at the diagram.	The Earth is shown in two different positions in its orbit.
3	Remember the key facts.	How long does it take for the Earth to go once around the Sun? How many months?
4	Work through the problem.	a) The answer could be in days, weeks, months or years. b) This is halfway through the orbit. Count on in months from position 1. c) Use the diagram to help you.
5	Check your answers.	

Try this

1 Draw a diagram in the space below to show how the Moon orbits the Earth.

2 How long does it take for the Moon to make a complete orbit of the Earth?

____

The Moon and the Sun are the same size. You can tell that this is true when you see them both in the sky.

Amy

3 Do you agree with Amy? Explain your answer.

____

____

Top tip

- Remember that at the same time the Earth is spinning on its axis, it is orbiting the Sun.

Day and night

To achieve 100 you need to:
★ explain how we get **day** and **night**
★ explain why the Sun appears to move across the sky.

What you need to know

- The Earth spinning on its **axis** causes day and night.
- It takes the Earth one day to spin once on its axis.
- The Sun looks as if it is moving across the sky during the day because the Earth is spinning.
- When the Sun is lower in the sky, shadows get longer. When the Sun is higher in the sky, shadows are shorter.
- Light travels in straight lines.
- We can use the information from a line graph to give an explanation of the results of an experiment.

Let's practise

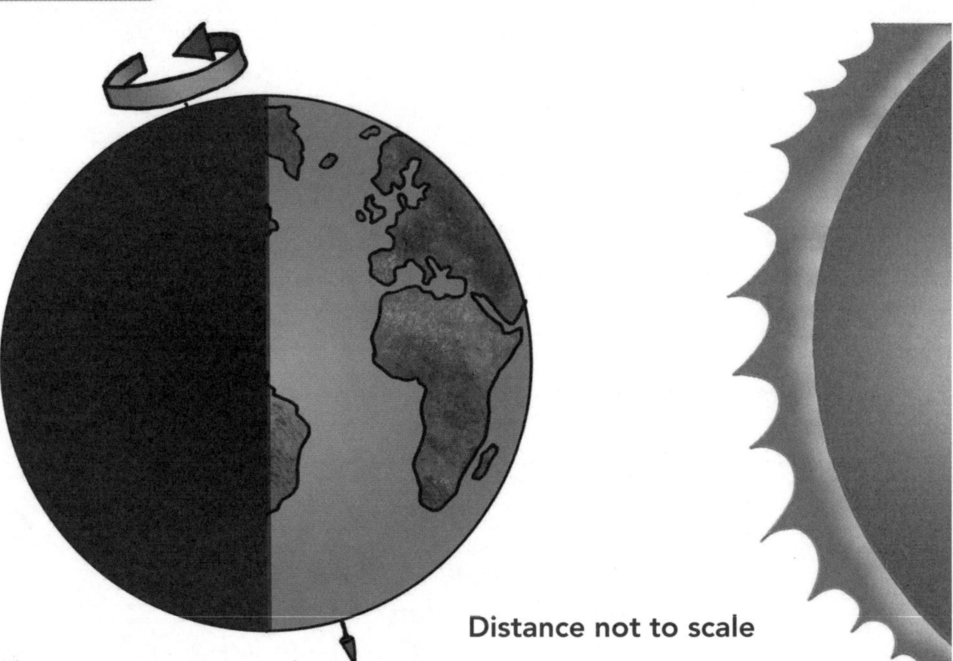

The diagram shows the Earth and the Sun.

a) Draw arrows on the diagram to show how the light travels from the Sun to give daylight on half of the Earth.

b) Use the arrows you have drawn to help you to explain why half of the Earth is always in darkness.

c) How long will it be before the Earth is in exactly this position again?

1	Read the question and read it again. What is it asking?	**It is asking you about day and night and what causes them.**
2	Look at the diagram.	**Look at how half of the Earth is in light and half is in darkness. What is causing this?**
3	Remember the key facts.	**How long does it take for the Earth to spin once on its axis?**
4	Work through the problem.	**a) Remember how light travels. b) Which parts of the Earth is the light reaching? c) Give the answer in hours.**
5	Check your answers.	

Try this

1 Describe how the position of the Sun appears to change in the sky over a day.

____

____

2 There is no daylight at the North Pole in the middle of winter. Explain why.

____

____

Top tip

- The position of the Sun in the sky depends on how the Earth spins on its axis.

Glossary

Adaptation The way a living thing changes to become better suited to its environment.

Circulatory system The heart and all the blood vessels that move the blood around the body.

Condense The change that happens when a gas or vapour turns into a liquid.

Conductor A material that lets heat, electricity or sound travel through it.

Digestive system The parts of the body where food is digested, starting with the mouth and ending with the anus.

Dissolving When a solid mixes with a liquid and becomes part of it.

Evaporate The change that happens when a liquid turns into vapour or a gas.

Evidence Something that supports an idea or conclusion.

Evolution The gradual process over time where living things change.

Fertilisation When the male and female sex cells fuse together.

Filtering When a liquid is passed through a material to separate out particles that do not dissolve.

Germination When a seed starts to grow.

Habitat The place where a plant or animal lives.

Impermeable When a material will not allow a liquid to pass through.

Inherited Characteristics passed on from parents or ancestors.

Insulator A material that does not let heat, electricity or sound travel through it.

Invertebrate An animal without a backbone.

Irreversible Something that cannot be undone (a change).

Microorganism A tiny organism that cannot be seen without a microscope.

Opaque Does not let the light through.

Permeable When a material allows a liquid to pass through.

Pitch How high or low a sound is.

Poles (of a magnet) The two ends of a magnet known as north and south.

Pollination When the male pollen is transferred from the male part of a flower to the female part.

Predator An animal that eats other animals.

Prey An animal that is eaten by other animals.

Producer A plant that produces the food at the start of a food chain.

Pulley A grooved wheel that is used to change the direction or the amount of force needed to move a load.

Pupil The opening in the centre of the eye where the light enters.

Reflect Sending or bouncing back (of light).

Repel When two similar poles of a magnet push each other away.

Reversible Something that can be undone (a change).

Sieving A way of separating out large particles.

Solution What is produced when a solid dissolves in a liquid.

Stamen The male reproductive part of a flower.

Stigma The female reproductive part of a flower.

Answers

Health and digestion

What happens to your food?

Let's practise 1 (page 8)

Plate b) contains a wider variety of foods; meat and vegetables; no fried food; not just carbohydrates.

Where does your food go?

Let's practise 2 (page 9)

a)

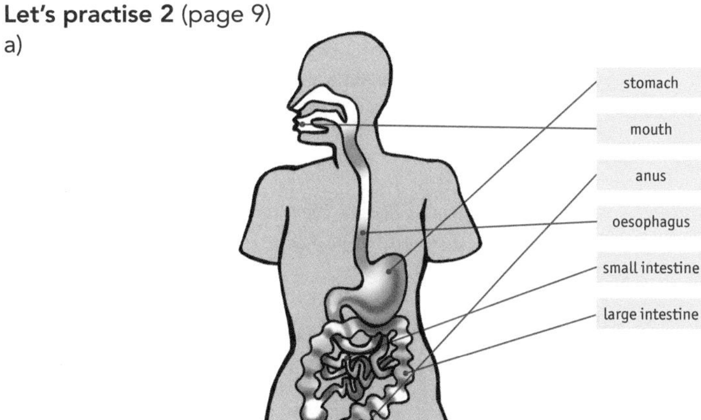

b) Neither child is correct: digestion neither starts nor ends in the stomach. Some children may indicate mouth and anus as start and end.

Looking after yourself

Let's practise 3 (page 10)

a) Less
b) Year 6
c) After Year 6 they do less sport as they get older. The bars on the bar chart go down after Year 6.

Try this (page 11)

1 Any reasonable answer, e.g. they have less time for sport.
2 Any reasonable answer, e.g. to make muscles/heart/lungs strong.
3 Any reasonable answer, e.g. so you have the right amounts of different food/vitamins for your body to function properly.
4 Exercise; not eating fatty foods; not being overweight, not smoking.

Skeletons

Let's practise (page 12)

a)

breast bone
thigh bone
pelvis
shoulder blade
spinal column
skull
ribs

b) Pelvis, thigh bone and spinal column

Try this (page 13)

1 Muscles often work in pairs. As one contracts and moves the bones, the other relaxes. In this way they make the joints bend and straighten. They are joined to bones by tendons.
2 Exercise; eat a calcium-rich diet, e.g. milk; not being overweight.
3 Many acceptable answers. Some simple creatures do not have skeletons, e.g. slugs and jellyfish. Some animals have external skeletons, e.g. crabs and woodlice.

Human development

Let's practise (page 14)

There are lots of possible answers, e.g.

a) Learn to walk
b) Grow taller
c) Voice breaks (boys)
d) Have babies
e) Grow shorter

Try this (page 14)

1 Humans only grow until they are adults. The age when they stop growing taller varies. Babies also grow inside the womb.

Teeth

Let's practise (page 15)

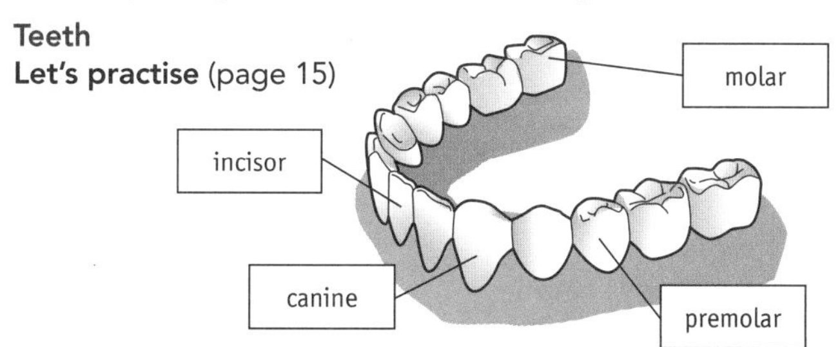

Incisors – for cutting food
Canines – for holding and tearing food
Premolars – for grinding and crushing food
Molars – for grinding and crushing food

Try this (page 15)

1 Brush regularly; use floss; visit dentist regularly; not too many sugary foods.

Food chains

Let's practise (page 16)

Leaf: producer; caterpillar: prey; thrush: prey and predator; cat: predator.

Try this (page 17)

1 There would be more leaves so the plants could find it easier to survive; there would be less for the thrushes to eat and this could affect their overall numbers; there could be fewer thrushes for the cats to eat.
2 grass → grasshopper → rat → owl
3 Any appropriate food chain with at least three parts starting with a plant. Arrows need to be drawn in the correct direction.
4 It is the only one that can use the energy from the Sun to make energy that can be used by the rest of the food chain.

Heart and circulation

Where does your blood go?

Let's practise (page 18)

a) Ben
b) Amy: the blood returns to the heart before going on to the rest of the body.
Sanjay: The blood is first pumped to the lungs then it goes back to the heart. It is then pumped to the rest of the body. It then goes back to the heart and then back to the lungs again.

A beating heart

Let's practise (page 19)

a) Two minutes
b) 130 bpm

c) She needs more energy as she exercises so her heart beats faster to get more blood to the lungs to collect oxygen.
d) The line on the graph starts and ends at the same heart rate.

Try this (page 19)
1 The heart is a muscle so exercise makes it stronger.
2 Eat a healthy diet; do not smoke.
3 Can be counted using pulse or a stethoscope or could be monitored by a heart/blood pressure machine.

Classification and keys
The key to identification
Let's practise 1 (page 20)
A = wasp, B = house fly, C = flea, D = ant

What fits where?
Let's practise 2 (page 21)
Amy: incorrect: reptiles lay eggs.
Sanjay: true.
Ben: incorrect: most amphibians have moist skin.

Try this (page 22)
1 There are many options for these answers, e.g.
mammals: cow; whale
birds: eagle; sparrow
reptiles: snake; lizard
amphibians: frog; toad
fish: shark; cod
2 It can help to identify organisms and to trace how they have evolved.
3 A suitable branching key drawn that enables leaves to be identified.

Inheritance
Let's practise (page 23)
a) Colour of fur/eyes; length of fur/tail/whiskers.
b) Small size may have been as a result of illness or lack of food.

Try this (page 23)
1 Many possible answers, e.g. height; hair/eye/skin colour; shape of face/eyes/nose.

Adaptation and change
Let's practise (page 24)
A sharp beak helps penguin catch fish; strong claws help it to grip on ice; fat body keeps it warm; strong flippers help it swim fast to catch fish.

Try this (page 25)
1 Global warming has reduced the sea ice, which is where the penguins breed and lay their eggs.
2 Fat bodies; thick feathers.
3 Many possible answers, e.g. rhinos due to poaching, orangutangs due to loss of habitat.

Investigating plants
What does what?
Let's practise 1 (page 26)
a) 1 leaves – where the plant makes its food (where photosynthesis takes place)
2 stem – water and nutrients are transported through this
3 flower – reproductive part
4 roots – take in water and anchor plant
b) The coloured water would gradually rise up the stem and change the colour of the flower. This shows how water is transported from the roots, up through the stem to the flowers and leaves.

Let's find out
Let's practise 2 (page 27)
a) Test three plants of the same type with all/no/half of the leaves cut off.
b) The plants should all have the same amount of water and be kept in the same place in similar pots with similar amounts of soil.

c) Plants are living things and may grow differently from each other.

Try this (page 27)
1 The plant with no leaves would probably die because it could not photosynthesise, although some children might suggest it would grow new leaves; the plant with half of its leaves would continue to grow and may grow new leaves.

The life cycle of flowering plants
Let's practise (page 28)
1 anther – contains pollen
2 stamen – male part (includes anther)
3 ovule – female part, which will become seed
4 stigma – where the pollen lands
5 petal – attracts pollinators if brightly coloured

Try this (page 29)
1 germination → flowering → pollination → fertilisation → seed formation → seed dispersal
2 Insects transfer pollen from one plant to another.
3 Brightly coloured petals; scent; nectar.
4 Two of: wind, water or animals.

Rocks and fossils
The rocks beneath your feet
Let's practise 1 (page 30)
a) Granite or marble because they are both hard and not permeable so would not get eroded.
b) They could put a drop of water on each rock and observed whether the water soaked in.

Fossils
Let's practise 2 (page 31)
1 sea lion – b allodesmus, 2 ammonite – a ammonoid,
3 trilobite – d trilobite, 4 coral – c horn coral

Where the worms live
Let's practise 3 (page 32)
From the top: twigs, leaves, etc: water; mud; tiny stones/gravel/sand; bigger stones.

Try this (page 33)
1 Several different ways, e.g. a skeleton is buried by mud, often in a river or the sea. More layers form on top and, over millions of years, the layer with the skeleton changes into rock under the pressure. Later movement of the rocks and erosion can expose the fossil.
2 They can give an indication of what extinct animals looked like.
3 The colour and skin of the animal.
4 There would be a lot of twigs and fallen leaves from the trees.
5 The rocks underneath could be different and this would make the soil different or there could be a different amount of organic materials.
6 Soil with lots of organic material for the worms to eat.

Solids, liquids and gases
How can you tell?
Let's practise 1 (page 34)
a) Amy: agree, property of a solid.
Sanjay: agree, although some solids, e.g. sand can also be poured.
Ben: disagree, a gas will spread out in any container.
b) Sand is made of tiny particles, each of which could be classed as a solid and you can pour it like a liquid. It will not form into drops.

Changing state
Let's practise 2 (page 35)
top row (left to right): melts, evaporates; bottom row (left to right): freezes, condenses.
1) 0 2) 100

Can you get it back?
Let's practise 3 (page 36)
a) It had started to melt/had melted. The temperature in the room was higher than 0°C so the ice began to melt.
b) She could return it to the freezer.

Try this (page 36)
1 The water in the air in the room condenses as it touches the cold glass.
2 Any two reasonable answers, e.g. melt chocolate/butter and then let it cool; let ice cream thaw and then refreeze it.

The water cycle
Let it rain
Let's practise 1 (page 37)
A evaporation; B condensation; C rain or precipitation; D collection

Evaporation
Let's practise 2 (page 38)

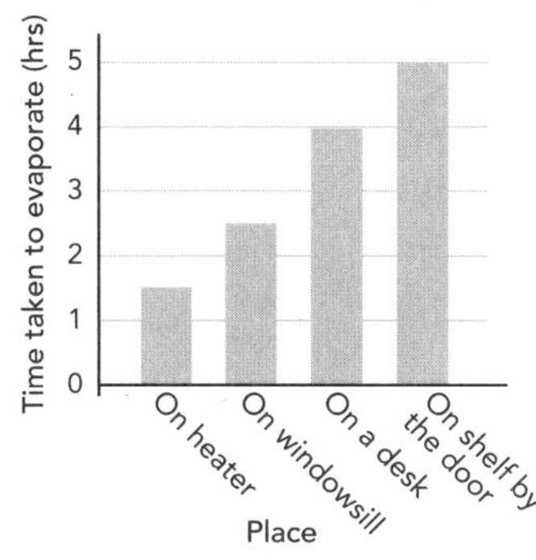

a) On the shelf because the water took longest to evaporate there.
b) The warmer the place, the quicker water evaporates/the colder the place the slower water evaporates.
c) Use the same amount of water and the same type of container.

Try this (page 39)
1 It could soak into the ground to become part of the ground water and then go into a river, which would flow into the sea.
2 Hail; sleet; snow
3 Somewhere warm, e.g. over a heater because the water from the coat would evaporate quickly.
4 On a sunny day the heat makes the water evaporate more quickly or on a cloudy day there is less heat so the water evaporates more slowly.

Reversible changes
Let's practise (page 40)
a) Information put into table:
Salt: disappeared, yes
Flour: water cloudy with some flour at bottom of cup, no
Sugar: mostly disappeared, yes
Sand: at bottom of cup, no
b) Salt and sugar.
c) You could filter them.
d) An investigation described to do with dissolving, e.g. time how long it takes salt to dissolve in different temperatures of water.

Try this (page 41)
1 Stir them.
2 A sieve.
3 The sugar dissolves in the tea. Melting is when something changes from a solid into a liquid.

Irreversible changes
Let's practise (page 42)
a) A (chemical) reaction has taken place.
b) An irreversible change. You cannot get the plaster of Paris back.
c) It will settle at the bottom of the container.
d) To see whether she could get the plaster of Paris back.

Try this (page 43)
1 In an irreversible change you cannot get the original materials back but in a reversible change you can.
2 Any two suitable answers, e.g. cooking a cake; frying an egg.
3 a) The vinegar and baking powder react together and a gas (carbon dioxide) is produced.
b) Irreversible.
4 It is not possible to get the original wood back. (The wood changes to ash/charcoal.)

Properties and uses of materials
Let's practise (page 44)
a) An appropriate investigation, e.g. putting some water on each fabric and seeing if it soaks through.
b) Same amount of water; same size piece of fabric.
c) Strength.

Try this (page 45)
1 a) It is hard; strong; will not melt or burn.
b) It conducts heat.
c) Wood or plastic because they are hard and do not conduct heat.
2 a) Allows heat to travel through it.
b) E.g. a metal saucepan allows the heat to travel quickly to the food and heats it up.

What uses electricity?
Let's practise (page 46)
a) Mains: washing machine, cooker, freezer;
Battery: remote control, torch, watch;
Both: laptop, radio.
b) Two of the following: never touch bare wires/anything electrical with wet hands; never poke anything into a plug; never climb an electric pole/pylon.

Circuits and circuit diagrams
Will it light?
Let's practise 1 (page 47)
Ben: The bulb will light because there is a complete circuit.
Mia: The bulb will not light because the wires are only attached to one end of the cell.

Symbols and how to use them
Let's practise 2 (page 48)
a)

Component	Symbol
cell	
bulb	
motor	M
buzzer	
open switch	
closed switch	

b)

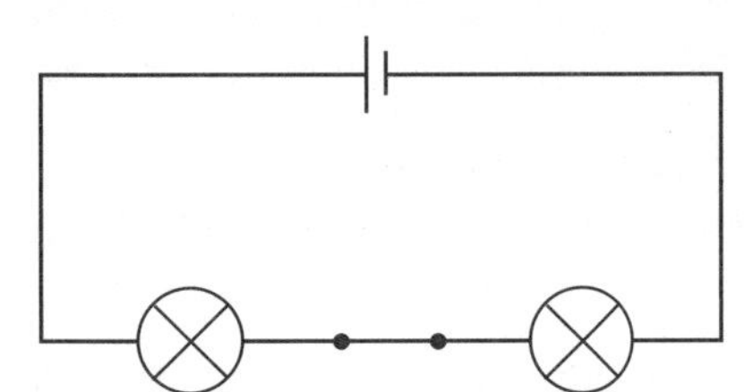

Try this (page 49)

1 a) Ring around wire not touching cell.
  b) Switch symbol drawn anywhere on circuit.
  c) Change the wires over either on the cell or the motor.

2

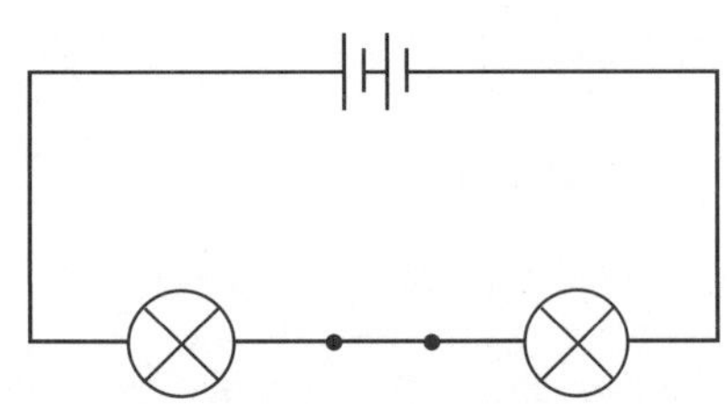

3 It will not make a difference to the bulb as a cell can be used either way.

Conductors and switches

Let's practise (page 50)

a) and b) An appropriate investigation that would enable children to find out which materials conduct electricity.

c) E.g. aluminium foil; metal paper clip.

Try this (page 51)

1 a) Neither will light.
  b) Neither will light.
  c) Neither will light.
  d) Both will light.

2 Copper is a conductor and so allows the electricity to flow through; plastic is an insulator and prevents anyone who touches it from getting a shock.

Changing circuits

Let's practise (page 52)

a) Jo: not true; they will be dimmer.
  Safi: true; there will be a higher voltage.
  Sam: not true; the voltage would be too high and the bulb would blow.

b) Correctly drawn circuit diagram with two cells.

Try this (page 53)

1 a) The more cells that are used, the louder the buzzer.
  b) They may have had the datalogger in a different place; they may have made a mistake in their readings.

Gravity and resistance

Let's practise (page 54)

a) Friction.

b) Yes because they may have made a mistake in their readings.

c) Highest, because it would take more force to make it move.

Try this (page 55)

1 Gravity is holding them on the Earth.

2 The larger piece has a larger surface area for the air to push against.

3 Appropriate example, e.g. moving parts of an engine.

Mechanisms

Let's practise (page 56)

a) Arrow pointing to top of small rock.

b) Use a longer lever.

c) It changes a small force over a big distance to a big force over a small distance.

Try this (page 57)

1 Use the plank and brick like a seesaw with the brick (pivot) near the teacher. Diagram drawn to show this.

2 Attach the things they want to lift to one end of the rope and pull on the other.

Magnetic materials

Let's practise (page 58)

a) Magnetic: steel paper clip, iron nail, steel scissors, steel saucepan.
  Non-magnetic: aluminium can, pencil, gold ring.

b) Extra suitable examples where magnetic objects have iron content (includes steel), and non-magnetic objects do not.

c) Keys can be made of different metals. Sometimes they can be magnetic and sometimes not.

Try this (page 59)

1 Description of a suitable investigation that would enable the children to compare magnets (all answers a)–d) will depend on choice of investigation).

2 Examples: in electric motors; for sorting scrap; in compasses.

Magnets

Let's practise (page 60)

a) They attract each other.

b) They repel each other.

c) They repel each other.

Try this (page 61)

1

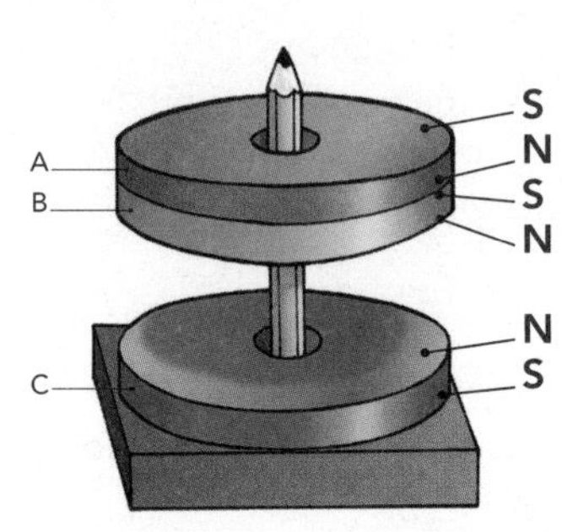

a) Two different poles are attracting each other.

b) Two similar poles are facing each other so they are repelling each other.

2

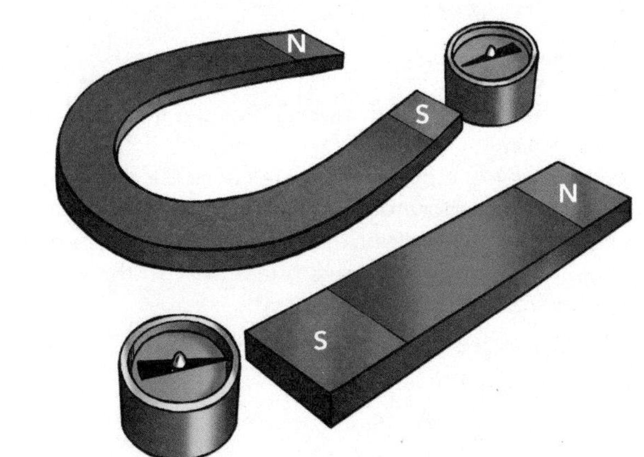

How we see

Let's practise (page 62)

Arrow drawn from bulb to picture; arrow drawn from picture to Benedict's eyes.

Try this (page 63)

1 a) The light bouncing off the object cannot get to the person's eyes in a straight line.
  b) Mirror drawn in a position so that the light can travel in a straight line directly to the mirror, bounce off it and straight to the person's eyes.

2 Diagram showing arrow coming from Sun to the Moon and arrow coming from Moon to person's eyes.

3 Don't look directly at the Sun; wear sunglasses or a hat.